TIME DOESN'T EXIST!

AND MANY OTHER THINGS …

(BIG BANG, BLACK MATTER, BLACK HOLES…)

Didier Viel

TIME DOESN'T EXIST!

AND MANY OTHER THINGS …

(BIG BANG, BLACK MATTER, BLACK HOLES …)

Didier Viel

It ain't what you don't know that gets you into trouble.
It's what you know for sure that just ain't so.

Mark Twain

Abstract

What we call time in physics is not a real entity in the universe. The real entities in the universe are matter and energy. Energy gives motion to matter or allows the transformation of matter, giving us the illusion of passing time or more precisely a perception of an "arrow of time".

The mistake about using time as a real entity in physics came first in the 18th century when L. Euler, in 1752, states that acceleration of a body is the derivative of velocity with time.

Instead, we will use d'Alembert definition of acceleration deduced from experimental results: $(\emptyset * dt) = dv$, where dt is not time differential as in differential calculation, but only a small constant of "duration", and ϕ the acceleration.

Einstein theory was to consider that time is a real entity in the universe and then space-time coordinates can describe the all universe.

Scientific advances are emerging and no one is aware of it since peer reviews prevents physicists to present new hypothesis which are not in the mold of Einstein General Relativity theory.

Nowadays physicists are dreaming to unify Einstein Relativity theory with Quantum Mechanics, despite all the evidence showing it is impossible to do so. Relativist followers are ignoring the fact that a new relativity theory has been proposed few decades ago which is fully compatible with Quantum Mechanics

Scientists have suggested new descriptions of the universe, with new hypothesis for Gravitation force and Inertia, and physics intelligentsia ignores all these scientific advances.

This book will also show why all the famous validation experiments of Einstein Relativity theory are untrue.

Table of contents

Prologue

Once upon a time …

That's how young children stories begin. The author is usually saying that the story takes place a very long time ago, and for this, he is featuring a story with fantasy characters living in a strange world.

It is a way to say to our little children that it was "before" and that they are living in the "present" which is different. It is much more referring to a long "duration" of time, but it is not quite the same "time" as what physicists are referring to today. In this book, we will show how "duration" and "time" are very different things in nature. Although duration of time is measured, it is a devious thing to give time a physical reality!

The "now" is referring to what we are doing. We know that we and our environment are under a constant evolution, and what was done a little duration before (a little while before) we can't do it again in the same manner, even a "duration" of a few seconds earlier. Sure, we can do it again but not exactly under the same conditions. Environment has evolved! We have evolved. This is the expression of the feeling of what we call the "arrow of time".

The notion of "time" existed as far as the 2nd Millennium before JC, but it was a way to measure "duration", which is very different from what physicists call "time" nowadays.

It was only from the 18th century that mathematicians allowed making a derivative of "time", which, as we will see, is the first mistake regarding the reality of time.

Later, physicists started using time as a parameter in all their equations for predicting events. After that, Einstein's famous Relativity theory come to associate space and time to create a universe with space-time as a metric.

It will be shown in this book that time must not be used as a real parameter in physics and why. With this new concept of non-existing time, it will be far easier to explain what we call the "arrow of time".

Introduction

Mark Twain said in one of his poem that the big problem people have, is not what they do not know, but what they strongly believe but is not true! This applies particularly in everyday actions, but not only!

This book shows how it applies to the physicist's community today. When being interviewed or when writing an article about physics of the universe, physicists often state that it is the only truth, instead of a working hypothesis to see how one can predict observations.

Science should follow a more cautious process such as: "I do not know for sure, but I am making this working hypothesis to see if I can correlate it to observations, and even if it is correlated to it, it should stay a hypothesis, since another theory, not yet known, can explain it as well". However, it is not what is being done.

Furthermore, when observing results, which do not obey a theory, physicists should say: "we should rethink this hypothetical theory as it might not be true". On the contrary, with conviction in their mind, they add some more hypotheses in a way that forces the theory to continue explaining what is being observed. Then there is not only one hypothesis but multiple and at the end they are so far from the starting hypothesis that they do not even know what they are doing.

However, the worst is yet to come when generation after generation of physicists pursued their researches while not even considering that their dear predecessors could have been on the false trail. In addition, when there is a lot of money involved, why can they stop their research based upon this theory?

For example, physics authorities today are considering the validity of Einstein Relativity theory done in so many experimentations, that there is no doubt this theory is the unique valid theory explaining the functioning of the universe and in particular gravitation force acting upon bodies in space. Then let us have a look to what it is about.

In this theory, the universe is supposed to have four dimensions in space and time and it is called a "space-time" universe. Gravitation force is supposed to come from the bending of the space-time universe by matter bodies in their vicinity, and to be proportional to their mass and inversely proportional to the square distance to the bodies. The given intensity of the force is the same as the Newtonian equations, except when gravitation is very high, say for example in the case of Mercury who is close to the sun.

Nevertheless, with this theory, physicists are not explaining the observation of the motion of the stars in Galaxies, the accelerated expansion of the universe and physics of the Big Bang. Rationale should have conducted them to say that, since it does not predict correctly motion of stars and Galaxies, maybe Einstein theory is untrue. It is not what they did and then they have to add new hypothesis.

For the first observation, they make a new hypothesis that the discrepancy of the motion of stars in Galaxies come from "invisible matter" in these Galaxies and they called it "dark matter". They do not know what it is, neither how it can interact with real matter to give the true motion inside the Galaxies. Then they supposed that there is a dark force between dark matter and real matter to give the correct motion of the Galaxies stars.

If you follow the reasoning, starting from a bend space-time, that is supposed to give rise to gravitation force to real matter bodies, it is then necessary to suppose that dark matter exists and "agglutinates" around Galaxies to give the correct gravitational motion inside the Galaxies by the effect of a dark force.

Then physicists came up with a new observation that all Galaxies are accelerating away from the solar system and from each other's! Physicists are saying that in Einstein bent space-time universe, there is a parameter call the "cosmological constant" which could explain the observations, but they do not know to what it corresponds in the physical world. We do not have to forget that the universe is real and then that we cannot have a parameter that we do not even know what it is!

Now they have a space-time universe, bent by matter bodies, fill somewhere close to the Galaxies with dark matter which is acting on real matter bodies by a dark force and then that between Galaxies there is something described by the cosmological constant in Einstein theory.

Next, physicists thought that since the Galaxies are accelerating away from each other, the universe was created by a "Big Bang". In order to get the physics of the Big Bang, physicists wanted to link Einstein theory with Quantum Mechanics, which explain the universe at atomic level. However, at this scale, in Einstein space-time theory, when space goes to zero value, one is getting an infinite singularity. To avoid this singularity, the ongoing hypothesis is that space-time cannot be smaller than some "grain", and it is called a "space-time loop". Now physicists are working on a "loop quantum gravity theory" to connect Einstein space-time Relativity and Quantum Mechanics.

Finally, we have a space-time universe made of "small loops", itself bent by matter bodies, filled somewhere close to the Galaxies with dark matter which is acting on real matter bodies by a dark force, and then that between

Galaxies there is something described by the cosmological constant in Einstein Relativity theory.

If you now come back to the starting point, you can see that you are far away from the initial hypothesis of a space-time universe and you have to wonder why the theory based on this unique hypothesis cannot respect by itself the observations. Before adjoining any new hypothesis, why do not rethink the initial hypothesis?

It is reasonable to go back to the starting point of a space-time universe and try to find what can be wrong with it.

Starting with space, the term space is a common name that covers several entities, but here we are referring on the part of the universe beyond our atmosphere. We know from observations that there are other matter bodies in the space universe. We know also from the Philosopher René Descartes, that tri-dimensional Cartesian coordinates can represent space.

Before Einstein Relativity theory, it was supposed that space was filled by an entity called ether. With this hypothesis, light was propagating through ether in some way like the sound is propagating through air. However, when an experiment aiming to validate this hypothesis failed, physicists thought that space is only void. Then in Einstein theory, space between real matter bodies was supposed to be empty. This experiment is the famous Michelson and Morley experiment, which we will discuss in details and see why it failed, and shows that it does not exclude the ether hypothesis.

New theories of the universe, based upon the existence of ether, filling entire space in place of void, can explain much more observations than Einstein Relativity theory do. After Einstein showed that one could not make difference between gravity and inertia, he thought that the force acting upon matter bodies exerting gravitation and inertia should come from the same entity. It would have been rational to think that ether is this entity but he did not go that far.

Then regarding space, is it emptiness or is it filled with ether?

If it is emptiness, then with Einstein theory of a bent space-time, what can be bent if space is empty!

There is not any proof of the reality of ether yet but anyway it cannot be very true that space is a void since in space, photons of light, electromagnetic waves and numerous particles are moving in it at or near the light speed. Therefore, space is probably not a void, but it is not really known.

Regarding time, it is even less obvious since numerous physicists are thinking about its reality and at present cannot have a clear explanation of what it is. It should be pointed out that everybody thinks that the hypothesis of Einstein space-time to describe the universe has been so much validated that it cannot be untrue. With such an assumption, it is impossible for them to think about time clearly in an impartial way.

The very valuable question to ask then is: "is what we call "time" a real physical entity"? In other words, is time is something that exists by itself or is it a byproduct of a real entity? We will see how to respond to this fundamental question and demonstrate that "time" is not a real physical entity in the universe and "duration" is effectively a byproduct of something.

When it is said that time does not really exist, it is needed to explain what gives rise to "the arrow of time". In doing so it will be explain by which real entity we have to replace the term "time" we are using.

When we say that time does not really exist, the obvious consequence is that Einstein's space-time universe does not exist as well. Then what would you say to people claiming that Einstein theory has been validated in so many experiments and that the GPS, people are using every day, is working by virtue of Einstein theory? In fact, explanations regarding how the GPS actually works, and why it does so without any corrections due to Einstein Relativity theory is given in the next chapters.

Then there is the need to go through every experiment claimed to validate Einstein theory in order to give clear explanations and verified arguments on the non-validation of this theory.

However, the starting point of Einstein Relativity theory was the need to have a transformation keeping the validity of all physical laws between two Galilean frames that are moving with a constant velocity between each other's. For example, it is not possible to transform the Maxwell wave equation from a rest frame to a moving frame with a constant velocity V, using Newton base calculations. That is to say that it is false to use the transformation: X= x+V*t, where V is the constant velocity along x-axis, and X and x are the coordinates on the co-linear x-axis of both frames. Then Einstein introduced a new transformation that keeps the wave equation true between the two frames in the space-time domain. This is the relativity principle and this was a revolution in physics at the beginning of the 20th century.

Nobody ever thought that there can be another transformation than Einstein's one that could exist and satisfy this same Relativity Principle. Another relativity theory has been proposed a few decades ago, that can also preserve all physical laws between moving frames with constant velocity between them and it is still ignored by the physics community. This new relativity theory does not need a space-time hypothesis to represent the universe, neither does it need real time entity to obtain that all physical laws are still valid between two moving frames.

However, it will be shown that this Relativity Principle is not quite respected by Einstein Relativity theory.

Also, Einstein's theory is the unique accepted theory to predict gravitation force in the universe. Nevertheless, there exist alternative theories proposing new hypothesis to predict gravitation force in the universe. It will be further explain that the whole universe may not be expanding as physicists are saying today based upon false interpretation of astronomical observations.

It is demonstrated that time is not a real entity and that it can be replaced by a real quantity that explain what is called the "arrow of time". In addition, a new Relativity theory has been proposed few decades ago. With this new theory, it results in consequences that are not as strange as Einstein's theory does.

An important part of this book dedicated to the analysis of most of the supposed validation experiments of Einstein theory show why they are not true or how they can be explained by other hypotheses.

Another part gives another vision of what space universe can be, with the presentation of some theories that explain gravitation force in a very different way from Einstein General Relativity theory. Those gravitation theories are so different from what one can expect gravitation force to be actually, that the reader may be completely surprised by these new theories.

Chapter1.

Time puzzle physicists

Throughout the ages, time was an essential question for physicists and even more at present. Nowadays, physicists are trying to unify Quantum Mechanics and Einstein General Relativity theory, to get an overall theory, which would represent completely the universe, from infinitely small dimensions to infinitely large, and time is a major element for this unification.

However, does time really exist? Why is there an "arrow of time" while the equations of physics are reversible with variable time?

This question of the real existence of time is a headache for scientists and philosophers, and a real nightmare for physicists.

Even today, there is no clear answer to what is called "time" and why there is an "arrow of time". A great number of physicists' claims that time do not exist, but it seems to be only a belief without a demonstration of its non-existence and an analysis of the consequences it can have on our knowledge of the universe.

Before proposing a new approach on the reality of existence of time in the following chapters, here is an overview of this issue through a bibliographic analysis. This will allow to move forward on the various points of view of this real existence. It is exhaustive enough to have a good comprehension of the issue.

First the article of Dan Romalo given during the 19th annual Conference of the NPA (Natural Philosophy Alliance), on the meaning of absolute time ([1]), concludes that it is not wise to consider time as a fundamental physical entity. His presentation is rather philosophical and unfortunately having made his conclusion, he didn't go through his ideas, since if time is not a real physical entity then by which entity to replace it?

French magazine "Pour la Science" dedicated a special magazine in 2010 on this subject ([2]), and this one establishes a good sample of the questions that puzzle scientists and of the answers they proposed to the real existence of time.

Professor Carlo Rovelli of CNRS (Centre National de la Recherche Scientifique) goes farther and thinks that it is necessary to free ourselves from the notion of time. He indicates that it is possible to formulate Newtonian mechanics, Quantum Mechanics and General Relativity without using temporal parameter. Indeed, Americans Bryce DeWitt and John Wheeler obtained an equation for quantum gravity where time does not appear! This result is obtained by applying the rules of Quantum Mechanics to the General Relativity Theory. Once again, Carlo Rovelli says we have to give up time on one side but continues to use the relativity theory on the other side, for which time is a necessity! Because if time does not exist what about space-time universe!

If one really wants to go to the end of this issue on the reality of time, then why not go to the logical conclusion and give up time as a real entity?

Physicist Etienne Klein of CEA (Commissariat à l'Energie Atomique) speaks about the "line" of time to represent his "flow", what raises the problem of the continuity of time instant by instant. He quotes an American scientist, Lee Smolin, who speaks about a «major abstract difficulty in particular when it is a question of unifying General Relativity Theory and Quantum Mechanics ". He envisages a thread of time simply bound to consciousness.

Philosopher Craig Callender concludes his long presentation on the fact that for him, time is as virtual as money can be in our transactions: " time allows us to relate physical phenomena's to each other's but time is a convenient invention which does not exist more in nature than money ". However, the author does not seem to raise himself the question: "and if Einstein General Relativity theory was false? »

Actually, scientists do not ask themselves this question, because as we shall see it, speed difference between reference frames requires having a rule of processing from a reference frame to the other one, which keeps the laws of physics true in every reference frame, taking into account the speed difference.

It is what is doing Einstein Relativity theory, but with disastrous consequences on the reality of phenomena's; for example, Craig Callender reminds us that with Einstein theory we lose the simultaneity of the events: «the determination of the events which occur at the same moment depends on the movement of the observer"! That is what is paradoxical and challenge the reason.

The new theory of relativity, which is proposed, can preserve the validity of physical laws between two moving reference frames and this without losing the simultaneity of events in the universe. Unfortunately, this theory proposed more than 30 years ago is totally untold!

It seems rational to have doubts on a theory, which does not keep simultaneity of events in two moving reference systems. An event has something absolute: it is or it is not, and it cannot be dependent on a clock, which would beat faster than another one.

Furthermore, as we will see in the next chapters, it is not the only argument, which arouses controversy on Einstein Relativity theory.

In this number of «Pour la Science ", Marc Lachièze-Rey, head of research in the laboratory "Astroparticle and Cosmology", evokes the disappearance of time in Relativity. He said: « Relativity of Einstein questions the usual properties of time. It is not possible any more to consider time as a universal and independent entity ". This assertion does not really question the real existence of time but evokes simply awareness different from what time represents. It is what indeed characterizes the theory of Relativity, which puts time equal to space. We can however object that unlike the space, time keeps a specificity, which is always characterized by an "arrow of time".

Marc Lachièze-Rey insists on the fact that with the General Relativity theory, the duration of a phenomenon depends on the observer (and of the velocity between two observers). This ensues from the application of this theory. He specifies that it is verified in numerous experiments, in particular in the showers of particles engendered by cosmic rays entering our atmosphere and the disintegration of unstable particles in particle accelerators. Did Marc Lachièze-Rey verify the validity of these experiments?

If yes, he would have realized that the first experiment is tarnished by a rough error. Is it an omission, a lie, or a fraud on the part of the authors?

As for the second experiment, a physicist who made a detailed analysis of it, shows that it is not validated by Einstein Relativity theory.

It is important to consult both chapters dedicated to these experiences, so fundamental for the validation of this relativity theory, to show the reader where the errors are, even if it is a bit difficult to understand.

Then Philippe Boulanger, former director of the French publication "Pour la Science" reminds us a number of paradoxes bound to the General Relativity theory, in particular the paradox of the twins to which we shall return after to show "the nonsense" of such a concept.

English physicist Julian Barbour is one of the pioneers to announce the non-existence of time based on the famous equation of Dewitt and Wheeler, but he admits that he was not able to go further in his announcement because of his lack of knowledge in mathematics! It is harmful for a physicist to rely completely upon mathematics. Yet mathematics is a tool and should not have the precedence on physics. Here is what Wikipedia quotes from the thought of Barbour: "numerous physicists criticize the fuzziness around its physical concepts. Statement: "time does not exist "is physically vague, but seemingly, philosophically clear. Philosophically, existence and time are not concepts the joint of which is obvious. On the other hand, to assert the existence of moments and nonexistence of time is at least paradoxical, especially as Barbour asserts that time is an illusion: thus nonexistence of time sends back to us to a realistic position. On the other hand, Barbour also asserts that time is not being useful for the physical theory, it is nothing, and it does not exist, otherwise as illusion".

We can also read in the magazine "Popular Science" what expresses Barbour: "have we live, we seem to move through a succession of Now's". However, Julian Barbour stopped on this conception of a continuation of moments to speak about "the non-existence of time", but he did not see what could replace or explain more exactly "this time which does not exist"!

In Quantum Mechanics, physicists question time. In the magazine "Discover Science for the curious", an entitled article "The time does not maybe exist" makes a reference to time measurements at microscopic level, realized by Ferenc Krausz. In his laboratory of the Max Planck Institute of Quantum Optics, he was able to measure the smallest interval of time of 100 attoseconds. To fix the ideas, 100 attoseconds is to one second what one second is to 300 million years. This experiment does not mean much for the non-existence of time otherwise by the fact that it becomes harder and harder to measure it the scales of atoms.

Again, in magazine "Pour la Science" there are very interesting articles on the perception of time by living cells (Patrice Bourgin, Etienne Chalet, Marie-Paule Felder-Schmittbuhl, Valérie Simonneaux). This is not the place to speak about these articles. Nevertheless, there is an interrogation bound to the paradox of the twins coming from the General Relativity theory which is this one: how the living cells of two people moving at different velocities, one close to the speed of light and the other one at rest, can do to age differently? This is a beautiful challenge for these specialists in biology.

An article on the perception of time by old traditional societies, by Eric Navet, shows that the spatial frame matters more than the temporal frame. In addition, route of their ethnic group replaces the "arrow of time".

To go further, on the site Elishean (³), physicists of the university of Geneva have just brought the proof that time does not exist in the microscopic world! The proof comes from correlating pairs of photons to give them a similar behavior, which seems not possible if time existed at this microscopic scale, since correlation occurred without time passing by!

In 2008 the FQXi Community Forum (⁴) proposed a platform on the subject "The Nature of Time Essay Contest". About 150 physicists or researchers answered this call and posted the "essays" on what the notion of time represent for them. It gathered an important quantity of very different opinions, which means that we are far from having an understanding on the reality of time right now.

George Snowdon is convinced that time does not exist and that it is only an illusion from us. He mentions that our timescale reference comes from the speed of the earth rotation on itself, which is 1670 km/h at the level of the equator. For him, absence of time makes universe much simpler to understand. There is neither past to visit nor future anywhere in the universe. In the absence of time, anything arrives simply.

Sandra E baron and Peter Wamai Wanjohi propose that the notion of time results from the motion of objects or from their evolution. As seen above the measurement of time is purely a convention.

Fotini Frameoupoulo complicates the situation by asserting that time really exists, but that space and gravity do not really exist!

Florian Girelli et al consider that time is very well defined by General Relativity theory as being a part of the space-time which is a dynamic object containing the gravitation.

Julian Barbour in his essay for FQXi essay Contest insists on the fact that time has nothing to do in physics. He quotes an interesting sentence of Ernst Mach (1883) "time is the year abstraction at which we arrive by means of the exchanges of things....".

He quotes then Isaac Newton who asserts that time is nothing other than the duration of things by the fact of the movement, which is the same idea.

In 1604 Johan Kepler showed that planets move around the sun (for a reference frame in the solar system) on ellipses, the sun of which is in one of the foyers of the ellipse, but the equation of the movement of planets around the sun does not contain time variable!

We use the movement of the rotation of the Earth on its axis to define our time, and we see that this definition is dependent from the reality of the universe. We could take another definition of time based on the movement of another planetary system. Julian Barbour demonstrates well that even if two clocks, one on the Earth and one on another planetary system, have not the same tick-tock, the events are perfectly synchronized. This means that in the universe, based on the movement of celestial bodies, they are all in correlation. There is no other clock than the clock of the universe. The reader can go verify the demonstration in the essay of Julian Barbour.

The reference made for the solar system is a good example of the human propensity to put itself at the center of the universe. The textbooks, which describe the movement of planets around the sun, are lacking humility because we do not say that it is a vision of an observer placed at the center of the sun. It is the same error made by our elder, which believed that the sun turned around the Earth because they put themselves at the center of the world. For an observer placed in the Milky Way, our Galaxy, the movement of the planets does not look like ellipses. Indeed, the sun turns itself around the center of the Galaxy at the speed of 965 000 km/h. Planets thus follow the sun in a movement which is difficult to represent, planets going sometimes faster and sometimes slower than the sun!

Artificial satellites Pioneers 10 and 11 are still the object of articles to try to solve an anomaly noticed on the two satellites. This anomaly is a gap of their orbits, when they leave the solar system with regard to their calculated trajectories.

Hypotheses, in extraordinary amount, very complicated and sophisticated to such an extent that scientists thought that it could be the starting point for new laws of physics! There is no limit to the creativity and there the scientists were very good.

Since then, the "Pioneer effect" affected other probes, as Near Earth Asteroide Rendez-vous in 2008. At the end of 2012, the scientists agreed on one of the hypotheses, which would be the reflectivity of the antennas of both probes Pioneer, but there is a simpler explanation, which is untold.

It reminds us of Galilee with the error of the year 1600! At that time, the Earth was thought to be at the center of the universe and fortunately, Galilee figured out the error. Since this date, it is common to represent the movements of the Earth and the planets with the sun as the reference frame. All our textbooks represent these movements by a rotation around the sun by ellipses, the sun being at one of the foyers of the ellipses. This description of the movement is only an apparent movement of the planets and it does not represent their real movement in the Galaxy. Then Pioneer effect is just miscalculation.

It is disturbing with regard to what is taught and obviously, it disturbs our physicists. Nevertheless, it is more rewarding to propose delirious ideas on the Pioneer anomaly than to return to notions of simpler flight mechanics to explain it.

Then how our planets are moving? The document in reference ([5]) gives a detailed description of the real motion of planets seen by an observer situated in our Galaxy.

This example of the anomalies of satellites Pioneer shows to what extent the scientists are attracted by exotic solutions, such as the multidimensional space, dark energy, dark matter, etc. to try to explain an anomaly, rather than being reasonable.

Let us return to FQXi essay Contest on the nature of time. Neil G. McCormick reminds us that according to R. Penrose ([6]) the laws of the thermodynamics have no direction of time. Basing itself on experiments, McCormick professes that all the chemical processes in nature progress according to their own logarithmic law function of time. According to this law, time can only go from zero to infinity.

Chris Kennedy asks himself an excellent question which is: "is there a physical mechanism which is responsible for what we perceive as time?". Unfortunately, he does not answer this question because he based his analysis upon the General Relativity theory.

This bibliographical analysis shows that researchers ask themselves the question of the real physical meaning of time but nothing significantly stands out from it.

The following conclusions can be resumed as follow:

- Julian Barbour sees the arrow of time as the pages of a book and although he does not believe in the real existence of time, he cannot give another definition,

- Carlo Rovelli explains that we can express the equations of Newtonian mechanics and General Relativity without time in these equations, but since he does not question the General Relativity theory, time has inevitably an existence through the notion of space-time universe,

- George Snowdon points out that there is maybe a difference between duration of event and time,

- Julian Barbour, Sandra E baron and Peter Wamai Wanjohi think that the notion of time comes from the movements of objects in the universe,

- Scientists try to unify General Relativity theory and Quantum Mechanics, but do not see any solutions at present and hope to get to it one day.

This is what we can retain from this bibliographic analysis of time and its reality. It is of importance to note that maybe the notion of time comes from the movement.

Chapter 2.

Why time does not really exist

It is well known by physicists that time can be removed in all equations in physics, meaning that all phenomena's in the universe are not due to what we call time. Then what can explain that physicists believed that time can be a real entity like it is for space?

In the universe, everything is moving or is transforming due to a process, Galaxies, stars, planets, rocks, etc. Each one is moving with its own velocity and/or is transforming with its own process. It is obvious that motion of matter bodies and processing inside matter bodies are due to energy inside or outside of these matter bodies.

To obtain motion of a matter body, a force is needed to be exerted on it. We know that a force will give acceleration to the matter body.

However, to relate increase of velocity and acceleration is not so obvious. When this problem occurs in the 18th century, it gives rise to discussions between great mathematicians.

It was d'Alembert (1717 – 1783) who expressed it in the following form, with dt as a duration [7]:

$$\emptyset * dt = dv \tag{1}$$

where ϕ * dt is "the quantity to which the increase of velocity dv is proportional", ϕ being what we call acceleration.

It is very important to stress here that "dt" is not a differential increase of time, like in differential calculation, but only a constant small "duration". Following d' Alembert, the expression ϕ * dt must not be separated in elements, since "dt" is a "duration" and that we cannot derive velocity v by a "duration".

This expression, established experimentally, can be considered as the best definition for the acceleration.

It is interesting to quote what d'Alembert states: "all the principle of mechanics can be deduced from the only consideration of movement".

Consider a body with velocity v, experiencing an increase in velocity (dv) on a distance interval (dx), v being the velocity on interval dx. The "duration" dt of this event is:

$$dt = dx/v \qquad (2)$$

Then the expression (1) becomes:

$$\emptyset * dx/v = dv \qquad (3)$$

Now x and v are real entity and we can multiply (3) by v and obtain:

$$\emptyset * dx = v * dv \qquad (4)$$

Even then, one can see that for a given interval dx, in order to have a given increase of velocity dv, the acceleration needed is proportional to velocity v.

Since:

$$v * dv = 1/2 * d(v2)$$

Then:

$$\emptyset * dx = d(1/2 * v^2) \qquad (5)$$

This is expressing that the acceleration on a small interval dx is equal to the variation of velocity square divided by two.

The force F = m * ϕ, can be expressed by:

$$F * dx = m * v * dv = d(1/2 * mv^2) \qquad (6)$$

A force exerted along a distance interval dx, gives rise to a variation of its kinetic energy.

And the expression of the force in differential form becomes:

$$F = d(1/2 * mv^2)/dx \qquad (7)$$

The expression of the accelerated force F, formulated by the x derivative of the kinetic energy Ek of the matter body, does not contain time.

$$E_k = 1/2 * mv^2 \qquad (8)$$

Then :

$$F = dE_k/dx \qquad (9)$$

One may make the objection from the above that the expression (3) is not correct, since the velocity is increasing during the interval dx, due to acceleration, and then duration dt cannot be equal to dx/v. A more exact derivation of duration dt can be done, taking into account the variation of v between x and x+dx. This calculation gives the same result as equation (2) when dv goes to zero.

Unfortunately, the expression physicist are using nowadays, attributed to L. Euler in 1752, expresses the accelerated force of a real matter body in function of time, as the derivative of the velocity v of the matter body versus time and multiplied by mass of the matter body m:

$$F = m * dv/dt \qquad (10)$$

It is interesting to show how this expression was established ([8]).

It comes from Galilée experiments with balls rolling on inclined surfaces. He obtains that position x of the ball obeyed to the law:

$$x = 1/2at^2 \qquad (11)$$

with a as the constant acceleration.

Increase in position between t and t+δt is given by:

$$\delta x = at\delta t + 1/2a\delta t^2 \tag{12}$$

In order to break free from δt, Leibnitz and Newton decided to make δt going to zero and obtain:

$$\delta x/\delta t = at \tag{13}$$

and then they define instantaneous velocity as:

$$v = dx/dt \tag{14}$$

and acceleration as the second derivative of x versus time.

However, it is more understandable that the acceleration force of a real matter body is the x derivative of its kinetic energy.

Since everything in the universe is moving or is transforming with a certain processing with its own speed, it is then obvious that what is real in the universe is not time but velocity! Moreover, since energy is the source for motion and/or processing of bodies, we can say that what is real in the universe is energy.

Physicists knows that there are two laws of conservation when there is an exchange of velocity or energy between external bodies and those are the impulse (mv) conservation law and the energy conservation law. From the above it is fully understandable.

Now since time does not really exist in the universe it is necessary to explain the meaning of what we call "time", and more over what gives rise to what we call the "arrow of time".

What we know for sure as a fact, is that our occupations, our way of living, etc... are driven by the earth rotation on itself. It is a practical way of knowing the speed of our actions and compares them to the moving earth. Then we call an earth revolution a day, we divide the day in 24 hours, and divide the hour in minutes and seconds.

Historically, man defines "time" only to measure "duration" of an activity relative to the moving earth. Then physicists define the parameter "time" as x/v, meaning it's the measure of the distance travelled divided by the speed. After that, they define a small interval of time by using $dt = dx/v$ meaning the small interval of time is a small interval of position due to velocity. Then one can come up with the velocity equal to dx/dt. But there is no reason to define velocity. Velocity is a real entity and we would have then to define a metric to compare velocities to each other's, in the same way as we do it for space with meter as a metric.

Wikipedia in its introduction on the definition of time ([9]) says the same: "there is no time measurement in the same way as there, for example, a measure of the electrical charge. In what follows it will be necessary to understand "measure of duration" instead of "time". The measure of the duration, which is the elapsed time between two events, bases itself on periodic phenomena (days, oscillation of a pendulum ...) or quantum (time of electronic transition in the atom for example)".

Wikipedia still gives the history of the measure of "duration" which goes back up as far that the IIth millennium before JC ([10]): "from the beginning of the IIth millennium, the Mesopotamians have counted in base 60 by using a numeration of position derived of the system of numeration of additive type and of mixed base of the Sumerians. This system is generally associated with the Babylonian civilization, which occupies the Mesopotamian South after -1800 and till the beginning AD.

This base crossed the centuries: it is found in the notation of angles in degrees (360 ° = 6 x 60 °) or in the cutting time (the 1 hour = 60 min = 60 seconds) of today".

The duration of an Earth rotation is used to define the metric for duration: one Earth rotation is defined as 24 hour, a second is defined by 1/3600 of an hour. At present, the reference of one second is established by other ways.

Thus the duration expresses itself by two entities which both have a physical reality, for example the circumference of the Earth at the equator (40 000 km) and the constant rotation speed of the Earth on itself at the equator. The Earth rotation speed at the equator comes from the definition of a day and an hour and then this speed is arbitrary; it gives an Earth speed rotation of 1 666 km/h.

It is usual to define the speed by:

$$v = \lim_{t \to t_0}\{(x(t) - x(t_0)/(t - t_0)\}$$ (15)

However, according to the definition of time with regard to the Earth rotation we have:

$$t - t_0 = \alpha(t) * \left(\frac{XT}{VT}\right), \tag{16}$$

with α(t) the proportion of the rotation of the earth between t0 and t.

When we make t goes to t_0, it means that we make α goes to zero. Thus, we have:

$$v = \lim_{a \to 0} \left\{ (x(\alpha) - x(0)) * \frac{VT}{\alpha XT} \right\} = \frac{VT}{XT} \lim_{a \to 0} \left\{ \frac{x(\alpha) - x(0)}{\alpha} \right\} \tag{17}$$

It means that to calculate the speed of a body, we compare the travel of the body in space divided by Earth speed on a fraction of the ground Earth rotation. This is nothing to do with an absolute time!

It is what we call time but in fact, it is a measure of "duration", and it is seen that it is only a way to compare what we are doing with the velocity of the earth! Other beings in the universe could do the same for their universe and it would be different from ours. If by any chance they came to visit us, they would measure duration for the duration between two events, and we will have only to know the rule from passing from one to the other means of measuring duration. Events in the universe are real and seen by anyone all the same.

What we call "time' is the comparison of the velocity of our actions compared to the moving Earth.

We can conclude from what precedes that both elements, which have a physical reality, are space, speed (energy), and not time.

What gives rise to the famous "arrow of time"?

A good example for this is the case of the broken glass. For some physicists there is a non-null probability that a broken glass can take its original constitution. This is because in all equations of physic, when we use time, time can go plus or minus. Then why in reality do we think that time can only take a plus value and so give rise to the "arrow of time"?

We have to think that since time is not a real entity then it is something else that gives the arrow of time.

Taking the case of a glass, making it is the result of a process perfectly defined and there is no way of doing it by chance. It needed to have a qualified personnel, adequate materials, and energy processing. But also when somebody let a glass fall down, since everything in the universe is moving or transforming, when the broken glass is on the floor, all universe keep moving and there is no way to get back to the state of the universe where it was before the glass fall down. It is certain that even by taking the pieces and with a trained personnel and a new process it will not give the original state and anyway the universe is not in the same state either. There is no chance to go back to the beginning of this situation.

Physicists say that it is due to the entropy of the system (the glass) which is rising. In thermodynamic for example, a system is said reversible if temperature of the sources and the system are identical. Then there is no exchange of energy and entropy is stable. If there is non-uniformity of pressure, temperature, density, then there is exchange of energy in the system and entropy is rising. In this case, it is said that the transition is irreversible.

It is clear for the glass being broken that there is exchange of energy and the entropy is rising.

We can say that the "arrow of time" is due to irreversibility of the states of the universe, itself due to exchange of energy. Everything in the universe is moving or changing states by consuming energy. In our daily activities, we are always moving and changing interior states by consuming energy.

The "arrow of time" is because everything in the universe is moving or transforming due to energy. Energy is pushing us in only one direction. There is no way to go back from one situation to another one. We can move in space from one point to another one and go back as we wanted to do, but doing that we must move and use energy, and the universe has changed too. Nevertheless, even if we stay still, not moving, our body is under transformation by many internal processes that are also using energy.

A very simple reasoning to represent the action of energy in the universe, that gives rise to the sensation of an arrow of time, is to look at water flowing out from a lake in a small torrent. Gravitation force acting on water molecules makes them flow down due to elevation change. Friction forces due to physical elements in the torrent then drive water flow. Velocities in the torrent are due to competition of all acting forces along the torrent course. Looking at the water flow could give the impression of the flow of time. In fact, there is no time passing but only energy at work giving the impression of time flow. Water molecules are subjected only to physical laws, and since the equations can be written without time, they are not aware of something we call time!

It was this sensation of "time passing" that drives people to realize one of the first clock to measure duration, using water flowing; it is called a clepsydra.

The movements of the Earth allow us in an obvious way to define what a day is and what a year is. The rest is only the division of this cosmological event. All the observers in the universe simultaneously observe the movement of the Earth whatever their speed of travel is and thus the simultaneity of the events is a fact.

We could also give a definition of time with regard to a frequency, since a frequency is the opposite of time. It is moreover the case when we use an atomic clock. According to Wikipedia, the atomic clock is used for «the conservation of International Atomic Time (TAI) and the delivery of the Universal Time Coordinated (UTC) who are reference timescales in our world ". Wikipedia gives the following explanation: «an atomic clock is a clock which uses the sustainability and the stability of the frequency of the electromagnetic radiation emitted by an electron during the passage from an energy level to another one. So the definition of time corresponds to the measure of a variation of a physical element (one might as well say of a movement) of our real space." We thus see that it is what we call the energy, which, being the mainspring of movements and of processes is at the source of what we perceive as time.

But the energy of matter, according to the famous formula of which it would be more correct to say that it is not attributed to Einstein but rather to Maxwell, is: $E = mc2$, where the matter internal energy is proportional to mass of the object multiplied by the square of light speed.

If we suppose that the speed of light is a constant in the universe, then the energy depends on matter and thus the only elements, which have a physical reality in the universe, would be matter (energy) and space!

By the way as indicated by Peter Kohut in his document ([11]), this expression of the energy internal $E = mc2$ was formulated by Maxwell in its theory of electromagnetism. Indeed according to him the momentum of a photon is $p = E/c$ and since $p = mc$, it occurs naturally $E = mc2$, without the use of any relativity theory. Peter Kohut explains how Einstein, who had knowledge of this law, wanted to demonstrate it from his relativity theory, made a mistake. The correct expression coming from Einstein relativity theory does not end in this formula!

It is now clear that there is no time passing but simply energy consuming and entropy rising, meaning we are always passing from one state to another one, and that process gives rise to the "arrow of time".

However, it is interesting to show that we use time as a parameter of duration to evaluate a mean velocity. For example for people running a 100-meter race, duration is a measure of the mean speed of the runner. Since it is not practical to evaluate the mean velocity during the running, by measuring the evolution of v, it is much easier to use duration:

duration t = 100m/mean(v).

It is also interesting to go back to the formula of the accelerated forces of a matter body:

$$F = mv \, dv/dx \tag{18}$$

One can see that as the velocity increases, the force to get the same amount of dv/dx increased. When running a 100-meter race, it is clear that the runner is rapidly encountering difficulty to increase its velocity since the efforts to produce becomes too large.

One can see that the definition of time we currently are using is a practical way of measuring velocity variation without knowing its variations. It is practical in our day-to-day activities, but we should not use it in physics.

For the expression of the acceleration force due to gravitation, we can express it also as:

$$F = d(mgh)/dh \qquad (19)$$

where g is the measured acceleration due to gravity and h the elevation with respect to a reference. The product mgh is known as the potential energy of a body in a gravitation field.

The expression F = dE/dx is then a general expression and means that the acceleration force is proportional to the variation of energy (kinetic and/or potential) along the direction of motion.

One can say from the above expressions that it is the variation of the energy in space, which is responsible of the motion of matter bodies in the universe.

It is also obvious that it is energy, which is responsible of the transformation of matter due to specific internal processing.

Time is not a real parameter in the universe. Energy change gives rise to the "arrow of time".

Chapter 3.

Instant (of time)? Past and future.

If time does not really exist in the universe then what is an instant ("instant of time")?

What we call an "instant of time" can be explained by a "configuration of the universe", meaning in this expression "configuration of the universe" that all entities in the universe are characterized by a position in space.

As an example: suppose that the universe is composed of particles. Every particle in the universe has one and only one position and no two particles can have the same position in space. For a given "configuration of the universe", all the particles are precisely defined in a unique position in space. It is what we can call an "instant" of the universe. Every event occurring in the universe can be related to a configuration of the universe and then to "an instant of time".

Supposing that everything is fixed, we have then what we call a photographic view of the universe. Let us suppose for the sake of the demonstration that the speed of light is infinite and then with our technology, a photographic view of the universe is a space representation in 2D of all the constituents of the universe. It is easy to understand that this representation is very complicated to express in simple terms. It is difficult to say that, for one of the fixed configuration, the Earth is precisely situated with respect to the sun, and that the other planets and all the Galaxies are precisely situated there and there …

Then to avoid this problematic we give a name to a particular configuration in space of the universe.

In the past, several calendars have been used. Each one was starting from a time zero defined from special events. For example ([12]) romans empire have used the founding of

Roma, in Japan it was the accession to the throne of a new ruler.

From this starting "configuration", astronomers were able to represent the universe and its evolution from this reference starting point.

Starting from its position in space, each particle has the ability to move, to change from one position to another one. Depending on physical factors like mass and energy, speed of particles is different and given by:

$v = sqrt(2*E/m)$, where E is the energy delivered to the mass m.

As previously mentioned, there is no time scale in this expression.

Then, after measuring position and velocity of matter bodies in the universe, astronomers are able to predict what can be a next configuration of the universe, from a previous one, when one of a specific chosen body has moved to a specific distance.

It is what we call the future. The future of the universe is already in gestation from a specific configuration of the universe.

For humankind future is much more difficult to apprehend since it depends on many parameters, not all known.

If a particular particle (or matter body) has a very constant speed, that we name v0, then the velocity of this moving particle (or matter body) could be used as a reference to compare velocities. For that, we should have defined a standard velocity as we have a standard meter for distance measurement.

However, what was done was rather define time by defining a "second" in reference to the earth rotation speed. Then we define duration being a quantity of seconds.

It is understandable that the above definition of duration is entirely arbitrary, but it is a convenient way to represent the state and the evolution of the universe from one configuration to another one.

However, the problem is that in physics duration became time and time became a real entity!

Going back to our photographic representation of the universe, there is another complication doing a photograph of the universe, since light, which carry the information of position and speed of matter bodies, is moving with a finite velocity. As noted in the French astronomical association publication in july/august 2017 ([13]), that since distances in the universe are huge, and speed of light is finite, that when we look at the sky or when we do a photograph of the sky, we think that we are looking at the stars in the past, in function of their distance from Earth! This is untrue, we see that stars cannot be at positions they should be due to light speed.

What we are doing in making a photograph is to superpose multi configurations of the space universe on a single medium. And then it is impossible for us to have the complete configuration of the space universe at an "instant of time"!

The representation of the universe by a photograph leads us to say that it is a time representation of the universe. It is only a representation on paper of multi layered configurations of the space universe.

Indeed, astrophysicists are able to relate the different instant of time by means of our calendar based on years.

To journey back to a specific "instant of time", it would be necessary, while going back, that the whole universe returns to this previous specific configuration.

Archeologists are researchers of past situations of configurations of the universe. It is possible then, since for specific reasons the evolution process of the "relics" of these past situations have suddenly strongly slowed down.

What we call the past is humankind memory and remnants of particular objects of past situations of the universe.

Chapter 4.

Relativity theories

Galileo had studied the idea of relativity almost three centuries earlier, when he stated the principle of relativity in 1632 ([14]). This principle says that the fundamental laws of physics are the same for all bodies in uniform motion.

Later in the 17th Century, Sir Isaac Newton also took the principle of relativity for granted, asserting that if his famous laws of motion held in one inertial frame, then they also held in a reference frame moving at a constant velocity relative to the first frame.

Necessity of a relativity theory occurs when it was clear that Newton way of calculation of Maxwell wave equation could not be transpose from a Galilean reference frame to another one when one is moving with a constant velocity and the other is at rest.

By applying Galilean transformation to electromagnetic Maxwell's wave equation, a different wave equation is obtained as the observer is in a fixed reference frame or in a mobile reference frame.

In fact, the equation of propagation of electric or magnetic field is in the form of the equation of d' Alembert as:

$$\left(\frac{\partial^2}{\partial x^2} - \frac{1}{c^2}\frac{\partial^2}{\partial t^2}\right)\varphi = 0 \tag{20}$$

where φ represents electric or magnetic field.

A Galilean transformation from one frame to another one moving with a constant speed ux along an x axis is given by :

$$x' = x - u_x t, \; y' = y, z' = z, t' = t \tag{21}$$

Using this Galilean transformation on the above d'Alembert equation leads to (^{15}):

$$\left(\frac{\partial^2}{\partial x'^2} - \frac{1}{c^2}\left(-u_x \frac{\partial}{\partial x'} + \frac{\partial}{\partial t'}\right)2\right)\varphi = \left(\frac{\partial^2}{\partial x'^2} - \frac{u_x^2}{c^2}\frac{\partial^2}{\partial x'^2} + 2\frac{u_x}{c}\frac{\partial^2}{\partial x'\partial t'} - \frac{1}{c^2}\frac{\partial^2}{\partial t'^2}\right)\varphi = 0 \tag{22}$$

This equation is entirely different from the equation at rest. It was then concluded that the Galilean transformation is untrue and that a new transformation should be discovered.

Einstein, with the help of Poincaré and Lorentz, proposed a transformation of coordinates that keep the wave Maxwell equation true between two Galilean reference frames in motion. It is called the co-variance of physical laws.

In this transformation, it is supposed that time do not pass by with the same rate in the two reference frame.

Using the following Lorentz transformation:

$$x' = \gamma(x - vt) \tag{23}$$

$$t' = \gamma\left(t - \frac{v}{c^2}x\right) \tag{24}$$

where: $\gamma = \frac{1}{\sqrt{1-\beta^2}}$, with $\beta = \frac{v}{c}$.

the d'Alembert equation stay the same in the two reference frames with relative constant velocity :

$$\frac{\partial}{\partial t^2} - c^2\frac{\partial^2}{\partial x^2} = \frac{\partial}{\partial t'^2} - c^2\frac{\partial^2}{\partial x'^2} \tag{25}$$

The idea admitted by academic science is that a theory is said relativist if it insures the covariance of the laws of physics from a reference frame at rest to another reference frame being in motion.

Since then, Einstein theory has been accepted as the true relativity theory. Einstein proposes two versions of relativity theories: Special Relativity and General Relativity. In simple word, General Relativity theory is supposed to explain gravity, with the idea that matter can bend the space-time universe defined by the Special Relativity theory.

Physicists never try to think that there is a possibility that another transformation transposing all physical laws from a reference frame to another in motion, could exist.

Young-Sea Huang has proposed a new relativity theory a few decades ago. Like Einstein Relativity theory, it is based on the principle of relativity and the constancy of the speed of light ([16]). The reader is encouraged to look in detail at this reference to understand the differences with Einstein theory.

Young Sea-Huang, in another article ([17]), shows that Einstein Relativity theory does not verify totally the covariance of the laws of physics. Indeed, in Einstein General Theory of relativity, all the laws of physics has to be expressed themselves under mathematical expression in the space-time coordinates system, what is not obviously the case of the law of conservation of the energy which does not express itself according to space-time coordinates. This only argument would be enough to conclude that Einstein General Relativity theory is not correct.

The basic principle of the transformation of the new relativity theory, transformation from a reference frame to another one with a constant velocity between them, is to do it in the velocity space rather than in a space-time coordinates. This transformation is called the differential Lorentz transformation and is transforming physical quantities instead of space-time coordinates, to make laws of nature form-invariant. It makes transformation of infinitesimal virtual displacements between the two reference frames using the speed of light as a reference constant velocity.

The differential Lorentz transformation is just like the Lorentz transformation, but in its differential form. It makes a great difference, since with this differential transformation, simultaneity of events is maintained. As said by the author, time and space are of Newtonian nature.

Moreover, a great challenge of Einstein theory actually is the fact that it has been proved that it is not compatible with the well-known Heisenberg uncertainty principle of Quantum Mechanic theory.

As quoted in the above reference, this new transformation ensures compatibility with Quantum theory : "In Quantum theory, according to the uncertainty principle of quantum mechanics, if one knows exact information in the space of velocity, then one cannot extract any information in the space of space-time coordinate, and vice versa. Based on the uncertainty principle, simultaneous transformation of both the space-time coordinates and velocities of events is untenable. The usual Lorentz transformation, which can simultaneously transform the exact values of the space-time coordinate and velocity, is incompatible with the uncertainty principle of quantum mechanics. The differential Lorentz transformation is a transformation in the velocity space; it is compatible with the uncertainty principle of quantum theory".

The strength of the relativists to show that the Einstein General Relativity theory is the true one is that they hurl that this one has been validated on very numerous experiments. It is thus necessary to deal with this point very seriously.

The first remark relativists are saying, is that the GPS (Global Positioning System) work only thanks to the relativist time corrections of the clocks of the various components of the GPS system. We shall see that this assertion is false.

There are much more serious experiments to argue of the validation of the theory. In particular the following experiments:

- time dilation of relativist particles such as atmospheric muons (particles which are created by cosmic rays interacting in our atmosphere), moving at speeds close to the speed of light,

- time dilation of muons in particle accelerators,

- deviation of star light by the Sun owed to the curvature of space-time,

- anomaly of Mercury périhélion.

And a few others experiments.

It is important to note at this stage that if a theory is validated on an experiment it does not mean automatically that it is valid. Indeed other theories can gives also a correct explanation and thus there is conflict.

There are so many experiments, which are supposed to make validation of Einstein theory, that now students cannot even thinks of going back and have a critical look at it to verify the validity of those experimentations. If you ask those students about it, they will think you are crazy. They will think that you are making a theory of the plot.

Then it is important to go through all the major experiences and begin with the GPS operation.

Chapter 5.

GPS and Einstein theory

Einstein and relativists worried a lot about the problem of synchronization of clocks. Indeed, with the General Relativity theory each reference frame have with it its own clock, which does not beat at the same speed as another one if both are in relative motion!

When physicists criticize this theory, it is sufficient to say: « the Einstein Relativity theory was fully validated and moreover it is used every day with the GPS, which, as everyone knows, could not work without Einstein Relativity theory". This kind of answer is very widespread among physicists who most probably did not try to study how it works.

More specifically let us look at the GPS operation and why, relativist corrections of time on the satellites segment are not used.

The GPS system was designed and is controlled by the American DOD (U. S Department of defense). This system is described in the document ([18]).

The basic principle consists in determining position of a ground receiver in 3D on the surface of the globe by a calculation of the distances separating the receiver with a constellation of satellites (we shall see that the system uses the information of 4 satellites).

A measure of distance sufficiently accurate for the purpose can be made only if the clocks of the satellites and the clock of the receiver are synchronized to the same time reference. Moreover, it is on this point that the relativity of time could intervene and it is with this argument that the relativists lean to assert that Einstein Relativity theory is validated by the operation of the GPS.

Indeed, supposing that this relativity theory is true, with an atomic clock on a satellite which would be left autonomous, then the drift of time of this one, due to relativity, compared with the time of the clock on the ground receiver, considering the significant speed of the satellite on its orbit, would be too important so that the receiver can't make a measure of distance precise enough for his purpose.

However, how the GPS system really works?

The system is based on a constellation of 24 satellites, which orbits around the Earth on 6 different orbits and a monitoring system on the ground constituted by stations around the globe and a chief station in Colorado. The chief station on the ground sends to every satellite the update of its precise position on its orbit and the universal time UTC (Universal Time Coordinated). After reception, every satellite emits a radio signal, which contains the information of its 3D orbit, its position on its orbit and time with regard to universal time received. Thus, there is no need for atomic clocks of the satellites to takes into account possible relativist corrections to avoid a drift of its clocks, since it is permanently corrected by the ground chief station. In fact, satellites would not really need an atomic clock.

What are the relativist corrections, which it would be necessary to take into account for the atomic clocks by every satellite?

The document in Neil Ashby's reference ([19]) clarifies, with supporting equations, the relativist corrections that should be taken into account by satellite atomic clock so that the measure of distance between the satellite and the receiver is compatible with the needs for geographical localization. These corrections allow minimizing the possible errors of a possible drift of the atomic clock.

It is important to note, that the possible relativist errors are not the only ones to take into account for a smooth running. The errors caused by the propagation of the electromagnetic wave into the earth ionosphere and by the refraction in the troposphere are bigger than the relativist supposed corrections ([20]). The way of freeing itself from these last errors due to the course in the atmosphere is to use two frequencies, which allows reducing these errors, but not totally.

GPS receivers do not possess an atomic clock. Its drift is much more important than the possible relativist corrections of the atomic clocks of the satellites.

Method of operating and extracting time for the receiver is very well clarified in a document which compares GPS and Galileo ([21]), the future European navigation system. In this document, it is explained clearly how the receiver, which does not possess an atomic clock can get back the universal time from the received signals. In particular, the messages that are sent by each satellite to the receiver contain the satellite clock time deviation (advance or delay) compared to Universal Time Coordinated.

GPS and Galileo do not use relativist corrections for the clocks synchronization of all the satellites. The GPS and/or Galileo are not the proof of the validity of the General Relativity theory since no relativist corrections are used.

Maybe Einstein theory is valid, but in no way the GPS (Global Positioning System) and/or Galileo constitutes a validation of this one!

Chapter 6.

Muons atmospheric time dilation experiment

This basic experiment is frequently highlighted as "the verification" of Einstein Relativity theory, by experimental measurement of time dilation of lifetime of unstable particles with high velocity. This validation is obtained by the measurement of the decay of relativist (having a speed close to the speed of light) unstable particles, which can be observed in the atmosphere, created during the arrival of cosmic rays at the top of our atmosphere.

This part is important and in spite of the complexity of the involved phenomena's. Even an unwary reader in this domain of unstable particles can manage to understand what takes place in our atmosphere with the arrival of cosmic rays at the top of it and of which we are going to describe what it is and especially what they entail.

Those who believe that this experiment demonstrates that time dilation predicted by Einstein Relativity theory is fully valid, take time (meaning doing a thorough reading) to analyze it in details.

According to Einstein Relativity theory, clock of a moving particle is slowed when its speed increases according to a law given by this theory and its universe is shortened. A moving particle having such a speed would have, with regard to a fixed frame, its life extended and thus the traveled distance seen of a fixed frame extended. This example is going to show that the devil is in the detail. Indeed voluntarily or not, the authors did so that the datum, which invalidates the experiment, is well supplied, but hardly discernible at first sight.

Accordingly, it is necessary to present in detail this experiment to understand where error hides.

After the discovery of unstable particles in the atmosphere, which moves at a speed close to the speed of light and which splits in a fraction of a second when they are at rest, it appeared to the upholders of Einstein Relativity theory, that the time of destruction of these particles, should be increased strikingly with regard to the one measured in laboratory at rest, due to their high speed. Thus if the measurement of the lifetime of these particles was achievable then it would be an excellent validation of the theory (if it was true and if it was the real reason!).

Unstable particles, created in the atmosphere by Galactic cosmic rays and solar energetic particles have sufficient energy to penetrate the geomagnetic field and the Earth atmosphere. When they enter into the atmosphere, they decayed into other particles by giving, in particular, unstable particles and so lead to a "shower" of particles.

 Protons mostly compose cosmic rays. The interaction between these particles and the core of atoms of the atmosphere produces unstable pions which split very quickly and produces, in particular, muons which have the peculiarity to have an average lifetime of 2,2 μs measured at rest.

So scientists have been considering that since muons speed are close to the speed of light, it could explain their presence at ground level by the slowing down of their internal clock.

A major experiment was led by David H. Frisch et James H. Smith of MIT, Cambridge, Massachusetts ([22]). This experiment taught everywhere in schools and universities as being one of the absolute proofs of the Einstein Special Relativity theory, is actually false.

Before presenting this experiment and criticizing it, we can wonder how an elementary particle constitutes in itself an inertial frame in which time would have its own existence connected to its relative speed with respect to another inertial frame connected to the Earth. But this subject looks like to be more philosophical than theoretical physics. So let us return to the experiment itself.

Here is the introductory summary of this experiment: «an experiment was realized to demonstrate that the relativist dilation of time is an important effect and this with relatively simple equipment. Among mesons-m1 (muons) reaching at the top of the Mount Washington (New

Hampshire), those who were selected had to have a speed between 0,9950c and 0,9954c. The number of those who survived to reach the sea level was measured at Cambridge (Massachusetts). The number waited without dilation of time was calculated from the distribution of the time of destruction of these mesons-m. (I.e. the life expectancy average measured as well in this experiment as in others and from the distance known for the descent. The factor observed by dilation of time is 8,8 ± 0,8 who must be compared with the factor calculated by effective dilation for mesons having these speeds in the geometry of our system of detection and being worth $1/(1 - v2/c2)1/2 = 8,4 ± 2$ ».

The authors realized an experiment in order to detect muons (or mesons) at two heights: 2000 m (at the top of Mount Washington) and at ground level. We can notice that the introductory sentence is significant of the state of mind of the authors who have only purpose to demonstrate the dilation of time through the appropriate slowing down of the clocks of muons due to their speed. Indeed their measures led them to a number of muons on the ground which corresponds exactly to the number of muons which would be surviving if they had been created at the height of 2000 m and if their clock was slowed down accordingly to the Einstein Relativity theory.

There is not, in the scientific literature, articles that analyses these results and we can wonder if somebody ever asked himself the question. It is surprising especially since this experiment is taught in all universities in the world.

This major experiment should be analyzed without preconceived ideas. If it is valid undoubtedly, then it is a demonstration of the validity of Einstein Special Relativity theory.

The proof the authors are looking for in this experiment is to show, that if there is still muons at ground level, then it is due to the Einstein Relativity theory. By the fact that the lifetime of the muons is increased due to their high speed according to Einstein theory, then muons can reach the earth floor, thing it could not do otherwise.

However, in order to be true they should have demonstrated that cosmic rays, giving rise to muons, could not go through Earth atmosphere under 2 km altitude.

To understand why there is still cosmic rays at ground level, and then can give rise to muons, it is necessary to have a look at the physics [23].

Electrically charged particles, like cosmic rays, have difficulties to enter the atmosphere as they tend to be deflected away by geomagnetic field of the Earth. The ability of a cosmic particle to penetrate the atmosphere depends upon its magnetic rigidity, which depends in particular to its momentum (speed). In addition, in each point of the atmosphere, there is a minimum magnetic rigidity necessary for the particle to pursue its trajectory down to the ground, called the "cut-off" rigidity. When the particle has a lesser rigidity than the cut-off rigidity of the atmosphere at a certain altitude, it will be deviate and will not attain ground level. Since the cut-off rigidity is inversely proportional to the square of geocentric radius, the cut-off rigidity rise when altitude decreases.

Then it is evident that close to ground level, only high velocity particles can attain these heights. It is what we can see on the graph fig. 20 of reference [24]. In this graph it is shown that between 2 km and the ground, the number of cosmic particles is almost constant!

Moreover, it is exactly what the authors measured! They mentioned they have measured 75000 cosmic rays per hour at the level of Mount Washington and 68400 cosmic rays per hour at ground level. The reader must take heed that the authors gave these data in a way such that the reader can't get it; data at Mount Washington are given in cosmic rays per hour whereas data at ground level are given in cosmic rays per seconds!

Knowing that in this range of altitudes, velocity of cosmic rays are the same at all altitudes between 2 km and the ground, due to their magnetic rigidity, then there should be about the same rate of muons created by the penetrating cosmic rays at these two altitudes; 2 km and ground level.

It is exactly what the authors measured!

Then why the author says that the presence of muons at ground level was due to Einstein Relativity theory?

The authors worried about the problem of penetrating cosmic rays in this range of altitudes, which as they say it could question totally the conclusions. They evacuate this problem by indicating that researches showed that there was no creation of muons in the portion of height between 2 km and the ground, but giving no reference! It is however the major point of the experiment. On internet, we can see that 100 % of articles on this experiment resume these conclusions without ever raising the elementary question of the validity of this assertion. It is dismaying.

This experiment is not a proof of Einstein Special Relativity theory.

Chapter 7.

Muons synchrotron life time experiment

Another interesting idea to validate Einstein Relativity theory consists in measuring the lifetime of muons circulating in a particle accelerator.

Nevertheless, the remark made in the previous chapter on the question: "is an elementary particle constitute an inertial frame with its own time?" is also valid for these experiments. It is necessary to wonder if this kind of experiment really establishes a test of the validity of Einstein Soecial Relativity theory.

In these experiments, we raise a subject difficult to comment because it is complex physics experimentation, which need strong skill in numerous domains of physics.

Four major experiments were led: in 1960, 1962 and 1969 at the CERN and in 1999 at Brookhaven. The main purpose of these experiments was to measure the anomaly of the magnetic moment of muons called "muons g-2 anomaly". We will not go into the details of this anomaly, which is not due to a possible relativistic effect. However, these experiments are showing seemingly an extension of the average muons lifetime when they are stored at high speed in a particle accelerator. However, there are questions about these measurements and there are other hypothesis to explain the results.

The authors of these experiments, all convinced by Einstein Relativity theory, did not analyze other possible eventualities. For them there should be no ambiguity on the proof of the validity of Einstein Relativity theory by these experiments.

But why the authors in the document of Francis's J.M. Farley on the experiences of the CERN ([25]), who are convinced in advance of the validity of the theory, uses unfriendly words for a person of the team who did not believe in it in advance and who was "fired" of the team. Why to use sentences of the style: "you see well that the drivers of car of formula 1 ageless faster, which goes to the sense of the theory of relativity of time"!

It is necessary to note that with muons, we are interested in particles, which according to Quantum Mechanics, should respect Heisenberg uncertainty principle, saying that we cannot know precisely at the same time their position and their speed. To know if the lifetime depends on velocity is it not necessary to know at the same instant position and speed? Is it not evidence that the principle of Heisenberg is false, and thus that the Quantum Mechanics is false if Einstein Relativity theory is true, and vice versa? Both theories are irreconcilable.

Francis's document J.M. Farley referenced above presents three experiments.

In the first case ((experiment of 1960 at the CERN), muons have a trajectory in the shape of solenoid and move longitudinally on a distance of 6 m. This device allows muons to make 2000 turns during their travel along the 6 m (corresponding to 2.5 km) and to have a time of storage from 2 to 8 µs. Let us remind that the purpose is to measure the precession of muons between the frequency of rotation of the circular trajectory of the muon and the frequency of rotation of its spin. It should be noted that this precession is independent from the speed of the muon as indicated by the author. On the other hand the interest of the relativist speed of muons is, if Einstein Relativity theory is valid and applies to the elementary particles, to be able to store muons longer and thus to make a better measurement of the precession.

The average lifetime of muons at rest being 2.2 μs, according to a law of exponential decay, the number of muons remaining later than 7.5 μs is about 3 %. It appears a low ratio, but if we receive in the sensor one muon per second, it does not seem so improbable if the number of muons at the source could be of the order of 100 muons per seconds. Without knowing the figure of the number of muons at the source, it is difficult to conclude that all muons had a slowing down of their clock. We can think also that this slowing down did not take place, and that considering the law of the life time at rest, there are always some who appear at the end at 7.5 μs .

In the second case of experiments at CERN (and at Brookhaven), muons are stored in a circular particles accelerator.
The authors assert that the average lifetime of muons

stored in the circular chamber is increased to a lifetime of 64 µs. The value of 64 µs is deducted from the temporal diminution of the number of decompositions of muons, diminution measured from the number of detected electrons coming from this decomposition of muons. If these electrons come from the decomposition of muons circulating in the accelerator, then there would be a significant increase of the lifetime of muons compared with the lifetime measured at rest.

The problem is that in a document of J. Bailey and al [26] which describe this experience, the low curve of figure 20 of this document shows in an obvious way that the average lifetime of muons injected is of the order of 2.5 µs [27]. In addition, we can notice that at the end of 4 µs there is almost no signal.

How is it possible that the signal increase again beyond 4 µs? Without additional explanation on the reason of this increase of the signal beyond 4 µs this increase remain uncertain.

Another curve given in the same document (figure 23) confirms a decrease of the order of 2.5 µs rather than 64 µs.

Eventually the experimenters supply a curve (figure 19) which allows to link the measures in the short times (from zero to 10 µs) with the one extrapolated from long times (from 20 to 189 µs). We can see a fast diminution from zero to 10 µs, probably owed to the destruction of muons according to the average life time of 2.5 µs, then the curve in the long times which flattens strongly beyond 20 µs. Difficult to know what really takes place beyond 20 µs with the data supplied in the document.

However, how the authors could find the number of 64 µs of average lifetime? It comes from the same figure 20, the upper curve that gives the decrease of intensity of the number of electrons resulting from a decomposition of muons, from 20 µs to 189 µs.

The author J.M. Farley attributes this curve to the density of muons presents in the accelerator and thus concludes in an increase of the lifetime due to Einstein Relativity theory. This result is consolidated by the calculated relativistic factor for the time dilation: $\gamma = 29$, (factor of slowing down of a clock bound to the particle according to the relativity theory) obtained from the distribution of radial position of muons in the accelerator.

Now assuming that it is the correct life time of muons which is measured in this experiment, the article of Young-Sea Huang in reference (^{28}) disputes the fact of obtaining the factor γ = 29,327. It is because J.M. Farley uses a circular reasoning to obtain this value: he makes the hypothesis that Einstein Relativity theory is valid to deduce a parameter he is using to obtain the value of γ = 29,327.

Starting from a un-relativist reasoning, Young-Sea Huang obtains, from the available data in Francis J. M. Farley article, a value of the factor γ between 6 and 11, what is far from the value of γ = 29 given by Einstein Relativity theory.

Francis J.M. Farley relate in the document in reference (^{29}) the second experience led at Brookhaven, in the same operating conditions as to the CERN. This experiment supplies the same extension of lifetime as to the CERN, but not the same value of the factor g-2 obtained at the CERN, this factor that was the objective n°1 of the experiment.

What should we learn of these experiments, highly complicated in their realization?

It seems factually that a curve gives an average lifetime in compliance with the average lifetime of muons at rest and another curve is supposed to give an average lifetime significantly increased with regard to their lifetime measured at rest, but of which the dependence on the factor γ of the Special Relativity theory is not validated.

There is much other doubt from the comments of Young-Sea Huang in the article, which seem justified.

Besides, comment of Francis J.M. Farley, in favor of Einstein Relativity theory, raises the question of the impartiality in the interpretation of the result of these experiences.

Finally it is necessary to ask another question which is: what can be the influence of the internal "energy" of muons circulating in the particle accelerator on their life time with regard to that measured while they are stopped in a laboratory test tube for a measure at rest?

Another aspect of these experiments was to verify the second postulate of the theory, which is the speed of light independent from the speed of the source.

For that purpose, the measure of the speed of photons gammas, stemming from the destruction of pions during flight, the speed of which was of 0,99975c, was made during the experience. The measure gave a speed equal to the speed c of light. It is not surprising, independently of the second postulate of Einstein Relativity theory, if we consider that the speed of photons depends on the medium in which they propagate independently of the speed of the emitter.

In conclusion, Francis J.M. Farley is convinced of the validity of Einstein theory but he made a circular reasoning by using this theory to determine a non-measurable essential parameter, which allowed him to prove that the theory is validated!

However, it seems that the lifetime of muons in the accelerator may be lengthened with regard to the one measured at rest.

This experiment remains fundamental and deserves more analyses to see if there is effectively an augmentation of the muons lifetime when they are stored in a circular accelerator.

Chapter 8.

Mercure perihelion anomaly / light deviation / Transverse Doppler / Hafele Keating's experiments

We have just seen some examples of experiments, which tried to validate Einstein Relativity theory for which time is a dimension of our universe, and which finally is questionable.

It exist many, many more experiments for which the upholders of Einstein Relativity theory assert that they validates it. We could examine them one after the other and show why the doubt is allowed on each of them. Numerous authors have made these analysis and these articles can be found in various publications and also in

Internet, but which are outside mediatized circuits. The list of these questionable experiments is very long and requires a lot of knowledge in physics and that is why they will not be handled here.

However we can mention two important experiments, which decided the scientific community to accept for true Einstein Relativity theory: the anomaly of the périhélion of Mercury and the deflection of the rays of light of a star by a gravitational field caused by the sun, two phenomena's predicted by the relativity theory. Later, two other experiments took place: the measure of the transverse Doppler and the experience of clocks around the Earth realized by Hafele and Keating, who are very famous among the current defenders of the Einstein theory.

Regarding the precession of Mercury, the simplest contesting would be to realize that the anomaly of the precession of Mercury is so low that any theory coming to predict it, would not constitute a very striking proof (Charles Lane Poor, on 1929, N. Rana, on 1987).

Indeed in Roger Rydin's presentation ([30]) we notice that the anomaly of observed precession is about 5600th arcsec

/ century which divides up into 5000th arcsec / century for the precession of the equinoxes, 530th arcsec / century for the disturbances by the other planets. The contribution of the Einstein Relativity theory would represent the non-explained complement to 43rd arcsec / century! Urbain Le Verrier had tried to explain the gap being lacking 43rd arcsec / century by introducing a new planet (Vulcain). But, finally a new planet not being discovered, this hypothesis was given up.

Albert Einstein by using his theory found a correction of 43rd arcsec / century, what allowed him to assert that his theory of relativity was valid. Furthermore, he predicted that rays of light must be bent by a gravitational mass such as the Sun.

Roger Rydin led an analysis of Albert Einstein's calculations and asserts that he made errors. A calculation without error would lead to a no correction at all for the Mercury precession.

Nainan Varghese gives another perspective ([31]) which would deserve a precise analysis. Indeed as indicated into chapter 1, it recalls that the calculations of Mercury perihelion is led by taking an elliptic orbits for the motion of the planets, while the real motion of planets is not a movement around the Sun! If the motion of planets is not an ellipse around the Sun, then there is no périhélion and there is no precession of this périhélion. The fact of representing the movement of planets around the sun by ellipses is a graphic representation of this movement, but has no real meaning in the physical world. Thus, a lot of time was lost by the fact that we always imagine ourselves in the center of the universe!

Einstein had also predicted that rays of light must be diverted by the curvature of space-time produced by mass objects. So when during an eclipse of the sun by the moon, it was observed that we could see a star, which being behind the sun on the axis Earth, Moon, Sun, should not be able to be seen, and that the measurement of the deviation of rays of light corresponded perfectly with the prediction made by Einstein, then physicists completely accepted the new theory.

However, as according to Einstein Relativity theory, it is the mass of the star, which creates, via the curvature of space-time, this deviation. Then it should be observed not only when rays of light touch a star but also when they pass at a certain distance of this star, for example at a distance of two or 3 diameters of the sun. As there is billions of stars, this phenomenon should be observed numerous times, and it is not the case.

Edward H. Dowdye demonstrated ([32]) that it is effectively the gravitational field of the sun, which is responsible for the curvature of rays of light but not like Einstein, predicted. This deviation of rays of light is caused by their interaction with the crown of plasma surrounding the sun, itself undergoing the gravitational field of the sun. It is this combined effect, which gives the deviation of rays of light. This plasma not being very vast around a star, this explains the fact that it is difficult to observe it. It is not curvature of space-time that can explain deviation of rays of light but the presence of the plasma surrounding stars as the sun!

Young-Sea-Huang, document in reference ([33]), criticizes the following two experiences: the measure of the transverse Doppler of a wave emitted by a transmitter in movement compared with the same wave emitted at rest and the experience of clocks circulating around the globe compared with a clock on the ground (Hafele and Keating).

For the first one, there are other demonstrations than Einstein Relativity theory, which explain the observed phenomenon. In fact, the demonstration made by the relativists leads to incoherence with the dilation time.

For the second, the experience consists of the comparison of two atomic clocks, one in a plane going round the world in a way, with others in a plane going round the world in opposite way and one stayed at the ground. The document in reference above supplied a very simple criticism, which is the imprecision of the data supplied by the authors in graphic form. We believe in it or we do not believe in it! However, an analysis of a gap, between a clock at the equator and a clock at the pole, according to Einstein Relativity theory, should be measurable, which is not the case apparently.

Chapter 9.

Big Bang theory and Black Hole

Big Bang theory is a cosmological model to describe the origin and the evolution of the universe ([34]).

The observation, made by Edwin Hubble in 1920 of the movement of the galaxies moving away from one another, contributed to strengthen the hypothesis that the entire universe is expanding, hypothesis, which had previously been envisaged.

This movement is measurable with the Doppler Effect, effect that moves the wavelengths of the emission spectra of Galaxies stars.

Edwin Hubble deducted a law from it, which stipulates that all the galaxies go away from us with a speed proportional to the Earth distance. Thus, the more they are distant from Earth, the more they go away fast from us. This law places us at the center of the universe! It is necessary to say also that this expansion of the universe can be predicted by Einstein Relativity theory, what was reassuring for their defenders.

Academic physicists are lacking objectivity, since it is necessary to make acrobatics hypothesis so that the Einstein Relativity theory can explain this phenomenon.

Shouldn't it be assumed that we are not at the center of the universe? Then if there is a red shift of the Doppler spectrum, all the more intense that the galaxy is away from us, it is not because it go faster, nor than it go away, but simply that the red shift is due to another phenomenon which is proportional to the distance Earth-galaxy!

A more rational explanation of this phenomenon is to make the hypothesis that photons undergo "a weakening" (tired light) during their motion in space in proportion to their distance to the Earth.

The article in reference ([35]) predicts the red shift of the cluster of galaxies called Corona Boralis by the theory of the weakening of the light. In this theory, photons are absorbed, then re-emitted by electrons of the plasma of the intergalactic space. Energy is transferred from the photon towards the electron and then the energy of the photon is reduced, as well as its frequency, and its wavelength is increased towards the red.

This is an explanation, which seems more reasonable as for the "centrality" of the human race.

Another proof, which is trying to make Einstein Relativity theory compatible with all the observed phenomena leads to more and more unrealistic nonsense.

Thus, exit the theory of the creation of the universe by the Big Bang. There is still work to come for the understanding of the creation of the universe.

Existence of black holes was discovered for the first time as a pure mathematical solution of the equation of the field of Einstein Relativity theory, called "Schwarzschild solution".

Unfortunately, as explained by Stephen J. Crothers in the note in reference ([36]), this solution is a false solution introduced by David Hilbert. The correct solution of Schwarzschild excludes any black hole! In the document in reference above the detailed explanation of this error is given.

Unfortunately, most of the astronomers and the astrophysicists, according to Crothers, do not know about this error and continues to use the bad solution. The original paper of Schwarzschild is indeed forgotten and almost lost for science.

Crothers also pointed out that the notion of dense star, the density of which is such as the light can extract with difficulty, was introduced by John Michell in 1784, from Newton's theory. To compare this notion of dense star with the black holes, it just takes a few easy steps, which were crossed by physicists, but this is another error. Contrary to the supposed black holes, in the dense stars the light can escape. They have no infinitely dense singular point and they can be observed. In his document, Stephen J. Crothers demonstrates that there are no black holes, nor no black bi-holes, in Nova Scorpii.

It is thanks to the error in the solution of the equation of Schwarzschild that the notion of black hole, predicted by Einstein Relativity theory, is going to continue. This demonstration still adds, if need to be, a stone to the invalidation of the General Relativity theory. The black holes have no demonstrated existence.

Chapter 10.

Ether / Michelson-Morley and Fizeau experiments / Sagnac experiment

Before Michelson & Morley experiment, physicists thought that space was filled with an entity called ether. Ether was supposed to be the support for the propagation of light.

By analogy in what takes place for the propagation of sound in a fluid, it was supposed that the speed of light had to change as it went down or went back up the current of the ether created by the motion of the Earth through this one.

They were two hypotheses considering the property of ether. The ether was supposed either to be stationary in space or either completely dragged by earth. The first case was the hypothesis used by Michelson and Morley, that is to say that the motion of the Earth in space created an "ether wind" at its surface.

Michelson and Morley experiments were then built to highlight the supposed «wind of ether» which had to modify time propagation of light between two beams in perpendicular directions. If the propagation of light through ether was like the propagation of sound in a fluid, then time propagation along the direction of this wind should be modified using Galilean combination of velocities. They were expecting to find differences of the measured speed of light in the two perpendicular directions.

For this purpose, they used an interferometer in order to superpose two beams of light propagating in perpendicular directions ([37]). Since there should be a time difference between the two beams, there should be interference fringes.

The experiments showed that there was no differences and that the hypothesis of stationary ether was not true and then Earth dragged that ether, if it existed.

During the same period, analysis of the results of another experiment complicated the analysis of Michelson and Morley. It was the Fizeau experiment ([38]).

The Fizeau experiment was to measure the effect of movement of a medium upon the speed of a light beam passing through it. Considering the above, it was supposed that the light was dragged by the medium. But the result there was a drag, but much lesser than what it was predicted.

To simplify, after several attempts to explain the results, Einstein pointed out the importance of the experiments for his special relativity theory, in which it corresponds to the relativistic velocity addition formula at small velocities.

Then what could be other explanations for those two experiments?

For the first one, Jean David ([39]), in his essay in reference, did an exact calculation by taking into account the fact that rays of light, as soon as they are emitted, they propagate in a straight line, while the various optical elements of the experimental set-up follow the movement of the Earth. In addition, he used Galilean combination of velocities of the light beam and the motion of Earth.

This exact calculation ends in a null result according to the Michelson and Morley experiment. What a surprise! Simple optical calculations using Galilean combination of velocities gives the exact result of Michelson and Morley.

Michelson and Morley null result is explained without the need of Einstein Relativity theory and this experiment do not invalid the possible existence of ether filling space.

We can thus conclude that there is no contraction of space and that this experiment cannot highlight the existence or not of ether.

These errors of interpretation were very much talked about and lead finally the scientists to accept Albert Einstein's interpretation. Michelson was overwhelmed at the fact that its experiments dedicated to reveal the existence of ether finally served to create the Einstein theory of relativity: " I created a monster "!

For Fizeau experiment it is demonstrated in reference ([40]) that there "is not partial dragging", as the special relativity asserts, but complete dragging of the light by moving medium. The decreases of the fringe shift in the Fizeau's two-beam interferometer is explained not with wrong Fresnel's aether drag hypothesis but with the phase deviations arising in the interfering beams, because of Doppler shift of the frequencies. Fizeau experiment does not prove but, on the contrary, refutes Einstein's theory of relativity".

Again, it is a false demonstration of the validity of Einstein Relativity theory.

It is now necessary to evoke the experiment of the French Georges Sagnac who resumed the experience of Michelson by fixing it to a rotating tray and by making so that both beams of light realize a complete tour, but in opposite directions. The experience consisted then to rotate the tray and measures interference fringes between the two beams. This experience was a success since G. Sagnac measures a difference of time of route, which he attributes to the presence of ether.

However, in this year of 1913, the theory of relativity had been born and Sagnac kept the door open to this one. Currently, it is not possible to discern if the effect noticed in this experience is due to relativity theory or to ether. Naturally, the upholders of the theory of relativity are not so generous because even if they admit that both hypotheses are valid at low-speed, what is the case in this particular case, they proclaim that there is a high-speed effect, which only the theory of relativity can explain.

We can note Pierre Spagnou's article ([41]), dityrambique on the theory of relativity which explains the Sagnac results, but to conclude too that it cannot be concluded between the effect of the ether and the relativity theory.

Actually, we notice that the phenomenon is due to time of route of both rays of light that are different because of the travel of the received sensor of the interferometer on the rotating tray. These are :

$$\Delta t1 = \frac{2\pi R}{(c-\omega R)} \text{ et } \Delta t2 = \frac{2\pi R}{(c+\omega R)}$$

with R the radius of rotation, c the speed of light.

By subtracting the two route times, we obtain an equal shift of interference fringes equal to: $\frac{4S\omega}{\lambda c}$, with S the internal surface in the route of the beams and λ the wavelength.

It is simply the addition of speeds in Newtonian physics, what is valid either if the ether exists or either at low speed if the theory of relativity is valid.

If this experiment does not allow to conclude, however the phase shift being proportional to the speed of rotation of the tray, which carries the transmitter and the receiver, this result gave rise to the development of lasers gyro meters embarked on planes, missiles or others, to determine in real time the angular speed in the plan of measure. With a combination of gyro-meters, it allows to know the real time trajectory of the moving body. This is a beautiful experiment, which led to a first-rate technical advance, and obviously, which the relativists hurry to claim that it is due to relativity theory while nothing is proved!

Finally, the physicists ended from the experiences of Michelson and Morley that the space was empty and that the electromagnetic waves propagate in empty space.

Moreover does the space is it so empty?

Peter Sujak ([42]) tells us all that there is per second in a cubic meter: " Thousands of protons, billion (10e12) of photons and neutrinos can be found per second in each cubic meter of the universe. " This is not so empty!

It is doubtless necessary to return to the notion that the space is not empty and that the way it is constituted is different from what thought the former scientists before the famous experience of Michelson and Morley. It would thus be necessary to re-analyze the conclusions in light of this notion.

However, of what is made the space which surrounds us? Hypotheses are proposed and it is important to say that with these hypotheses it explains more phenomena's observed in the universe than Einstein Relativity theory.

Indeed, both basic phenomena's in our existence that are the inertia force and the gravitational force are not really explained by Einstein Relativity theory.

For example how to explain that, the curvature of space-time can provoke any strength of attraction on the scale of the size of an apple.

For the gravitational force of celestial objects as the Sun and the Earth, the relativists represent the curvature of space-time by a paper sheet with at its center the Sun, which comes to deform the paper sheet and a planet which turning around the Sun experiences a deviation because of the "slope" of the paper sheet. What error! If the sheet sinks because of the presence of the Sun, it is because of its weight! In addition, from where comes the weight of the Sun on the sheet? Besides, the space is not bi dimensional but tri dimensional! Then how can we believe such a silly thing with such an incorrect example.

For inertial force, relativists recognize that Einstein Relativity theory cannot explain it. Then why not deal with the new theories, which give a rational explanation for these two forces, gravity and inertia? Moreover, these new theories are able to explain most of the observable phenomena's of the universe and this without dark matter and dark force!

Is it because there are several theories, which we cannot decide which one is the right one that physicists denies them? It is safer having only one theory to taught to students, what gives an impression of truth, while if there is several it could be thought that physicists wade and finally do not know much, what unfortunately seems to be the case. In any case, if we remain where we are now, we will have no chance to progress in the right direction.

Chapter 11.

Aging and Einstein Relativity theory

In 1911 Paul Langevin formulated what is called the twin paradox: let us imagine two twins, among whom one makes a spatial journey at a speed close to the speed of light whereas his brother stays quietly on Earth. When he returns, the twin traveler will have less aged than his brother will, because aboard the vessel, time will have more slowly passed by than on the planet.

This is an absurdity and a complete contradiction since the relativity principle states that the laws of physics stays equally true in any reference frame at rest or in motion.

Contradiction is to affirm that on one side "time" is dependent on the relative motion and on the other side that the laws of physics stays true.

It can be demonstrated by analysis on what the life behavior of a person in a spaceship moving at high speed is dependent.

The environment of its vessel must be very close to the one who stayed on Earth with matter of life survival and his physical activity has to be the same of when he is on Earth. We know also that in any condition, a person has to make physical exercise to keep in shape and we just stated that the laws of physics are the same than on Earth.

Let us look at the heart functioning in relation to physical activity.

Regulation of the heart is made through the blood pressure of the vessels of the blood circulation, via pressure sensors situated at the level of the heart and of nervous organs transmitters, decision-makers, then actors. The blood pressure increase when the muscles cells in the body need more oxygen to respond to an increase of activity of the muscles. The need of the cells comes from physiological laws and environmental conditions. The cells are acting along their own biological needs.

It ensues from it that when the blood pressure increases the heart rhythm increases. The blood pressure thus increases when the body demands a supplement of oxygen during the passage inactive – active (physical exercise for example). At the end of a number of movements with a constant rhythm, the beatings of the heart, having undergone an increase of the heart rhythm, stabilize with a certain rhythm. If the exercise continues over a long period, this rhythm remains constant, due to the cells constant oxygen demand. Then when the person returns to its normal activity, the rhythm of the heart slows down until a new balance.

All this to show that whatever of what we call the tick-tock of a clock, our heart is functioning according to a closed loop mode driven only by biological laws.

In a spaceship, whatever its space-time would be, the laws of physics are kept true according to the principle of relativity. The physiological laws also are kept true because they depend on the laws of physics. The environmental conditions inside the spaceship are identical to that on Earth to allow human beings to live. Thus, a human being inside a spaceship is subjected to the same rules of regulation of his heart rhythm as on Earth. Qualitatively we often say that the life expectancy of a person depends on the state of its heart and on a number of cardiac pulsations supposed to be maximal. Then this is a simple argument to show that aging is not slowing when a person is moving at high speed, in contradiction with the time dilation of Einstein theory.

It is amazing, while scientists do not even know the very root cause of aging, mathematic equation of Einstein Relativity theory is able to explain human aging! Why people believing in Einstein theory could be the best people able to explain aging?

This example showed that it is absurd to say that the laws of physics remain true in all reference frames and that there is time dilation when a reference frame is moving at high velocity, then different aging of people.

Chapter 12.

How to build a physical theory of the universe

Since Einstein, theoretical physics became the property of mathematicians and this phenomenon favor the exclusion of physicists who would possibly not have sufficient mathematical level to analyze the equations of Einstein Relativity theory. It is easy indeed, for the upholders of the relativity theory, to say that if people criticize the theory, it is because they understand nothing and then they do not believe in it. Moreover, for them it would be necessary to believe in it only because it was validated by numerous experiences. The trouble is that this argument is based only on claims without irrefutable proofs and which are explicitly denounced from now on.

In physics, mathematics is used "to simulate" a physical phenomenon when the laws of behavior of a new phenomenon has been discovered and understood. For example, the statistical formalism applies well to the modelling of gases because we established that a gas consists of molecules free of their movement and this with a random behavior. The kinetic theory of gases allows modeling the behavior of gases from the individual movement of molecules that constitute it, and allows determining the macroscopic values such as temperature and pressure of the gas.

For example, the equations of Maxwell, describing the propagation of electromagnetic waves, were established from four equations of behavior of the electric field and the magnetic field. Each of it was formulated according to observations in laboratory, and not the opposite. By combining these equations, Maxwell ended in the equation of propagation of the electromagnetic field. Even if today these equations are disputed given that they are approximations, the Maxwell equation is sufficient to describe what we observe. It is obvious that it was established from laws of behavior deducted from experiences of physics and not vice versa.

By the way, we can note that it is this equation, among others, that guided Einstein in the elaboration of his relativity theory, because they have to remain invariant in a change of coordinates between two reference frames in motion. It is this property, which gave its legitimacy to Einstein Relativity theory.

However, why trust mathematics formulas to predict physics phenomena's? Is it not a fatal error to proceed in this way because why we would be capable of finding ex-abrupto a mathematical equation, which explains the functioning of the all universe?

When we established a mathematical equation, how do we do to explain the reality of the phenomena's taking place in the universe?

Reality for gravitation force is how to explain that elementary particles constituting objects are attracted towards each other's? In Einstein Relativity theory, how to explain that particles are attracted to each other's, due to the curvature of the space-time they are experiencing?

Reality for inertia force, is not to assert a principle of equivalence between gravitational mass and inertial mass.

It explains absolutely nothing; it evacuates simply the problem of the reality of the phenomenon.

In 18th century, d'Alembert was wiser than scientists of present century were. He points out that matter body are subject to three very different situations: body at rest, in uniform movement, in uniformly slowing down or accelerating movement, which led to three sorts of effects. But, because the body is not different in these three situations, he concludes that it is the same cause which is differently applied on the body, this same cause which cannot be attributed to the body itself. For him the common cause of these three effects is the acceleration. Unfortunately, he was not able to go farther and explains what creates the acceleration.

Until now, consideration was given at the aspect of mathematical prediction of the physics of the universe and we demonstrated why Einstein Relativity theory could not be any more considered as valid.

A new theory of relativity is proposed by Young Sea-Huang who possesses robust arguments to replace Einstein theory, and opens up new horizons.

Is the complete media blackout, caused for more than a hundred years by Einstein Relativity theory, finally going to give way and move forward to the understanding of the universe?

It will be needed to go back hundred years ago!

Now it is necessary to consider the aspect of physical description of the universe. These theories are trying to explain the whys and the hows of the functioning of the universe unlike the mathematical description to represent the universe by mathematics formulas.

Before approaching the new physical theories of the universe, it is important to evoke Quantum Mechanics. Quantum Mechanics is the branch of the physics which describes the fundamental phenomena's at atomic level and subatomic scale ([43]).

Though physical interpretation of Quantum Mechanics is not making unanimity within scientific community ([44]). Indeed, there are not less than about ten interpretations of the phenomena's that are described by Quantum Mechanics. Thus, there is no consensus, but at least there is intellectual confrontation, what allows moving forward without risks to go in a dead end.

Besides, one of the problems with Quantum Mechanics is that it was born after the recognition of Einstein Relativity theory and after the experience of Michelson and Morley, experience which swept the hypothesis of a medium filling the space for the propagation of light waves, the so-called "ether".

Accordingly, Quantum Mechanics had to develop in this context with space-time and an empty space of any substance.

For quantum physicists, particles constituents of atoms are in an empty space and those particles of the atom nucleus (protons and neutrons) are maintained between them thanks to the said "strong interaction force". However, from where can come this force?

Quantum Mechanics tells us that this force is carried by particles, which are called gluons, in the same way as the electromagnetic force is carried by photons. Yes but from where comes the energy which contributes to give energy to gluons? It is the question, since in an empty space it is difficult to imagine from where this energy can come from.

For electromagnetic force, it is said to be a weak interaction, carried by photons, which propagates in empty space, their energy coming apparently from interactions within matter. Nevertheless, for gluons it is different since they act on matter.

It is these kinds of questions the new theories of physics of the universe are trying to answer.

If we look in the Medias, we cannot find any information about alternative theories. Given, any other theory not based on Einstein Relativity theory, based on space-time, cannot find its place in the scientific community. What is causing the problem of this strange marginalization of these theories and their authors?

It is necessary to understand that "academic" theoretical physics, is practiced by physicists with mathematical formation. Einstein Relativity theory is of a strong mathematical complexity, what made say (and write) that there were only four people into the world able of understand these equations!

If we do not know how to treat complicated equations, then we are not credible and we are neither listened to, nor read. Furthermore, if the ideas do not contribute to completely silly theories, then there is any luck to be listened to.

The category of the "dissident" physicists who propose alternative theories, favors the representation of the world by real entities. In their presentation of their theories, there is no mathematical equation and thus how could mathematicians believe in it?

Who is wrong, who is right? In order to deal with this issue, the first category of physicists has to condescend to deal with the theories of the second.

How does a theory of physics have to build itself? This question, however simple it is, deserves to be developed. The mathematical modelling of the physical phenomena's, when we understand the functioning of such or such phenomenon, with laws validated by experience, is useful to make forecasts of behavior in situations different from the experience.

However, the fact of using a mathematical formulation ex-abrupto, then validating this mathematical formulation a posteriori based on observations can lead to do errors. The article in reference (^{45}) gives an excellent explanation of errors committed at present in the field of theoretical physics.

The general idea of the introduction given by Nainan Varghese in his book is illustrates here with the change in hypothesis of Einstein Relativity theory confronted with contradictory observations.

"A theory has to base itself on hypotheses. When these hypotheses are translated into mathematical equations, the game consists in determining the consequences of these hypotheses and to verify if they allow explaining what we observe. Generally, plan events which did not occur yet and which we discover during a new experiment, it is similar to the magic and the theory is largely convincing there. It is what occurred with Einstein Relativity theory, when it was observed during an eclipse of the sun by the moon, the radiation coming from a star hidden by the sun and the theory of which had planned this effect by the curvature of space-time ".

Actually, this phenomenon can be understandable in quite a different way!

Let us continue with the text: "so if a theory showed itself true in a domain, physicists judge that the theory is valid and pursue its use to explain a maximum of physical phenomena. Generally, we fall on a new phenomenon, which we do not manage to explain, and then no problem we change, we modify, and we add hypotheses. This process is pursued numerous times and can lead to a set of very complicated hypotheses. For example, we observed that in the distant galaxies, the theory of the gravitation given by the curvature of space-time does not allow to explain what is observed."

Here is what we find in Wikipedia: «The gravitation is described by the General Relativity today, according to which the masses (or more exactly energy and streams of energy) bend the space-time. The Universe containing a multitude of masses, the space-time has a very irregular shape. However, on a very large scale, it seems that the Universe is very homogeneous (as a sheet steel of metal can seem rather flat seen from afar, even if it is quite dented seen closely). We can then ask ourselves the question of the global geometry of space-time. It is an experimental question that can be proved by astronomical observations. Indeed, the distribution of rays of light is sensitive to the global curvature of space-time, and the observation of distant objects allows determining it. One find that the space is flat, and then that the global curvature of the Universe is null! In a theoretical way, it turns out that this curvature is connected with the total density of material and energy. The flatness of the space-time allows calculating with a good precision the density of mass-energy of the Universe: it corresponds to 5.7 atoms of hydrogen by m3. The density, which gives the flat space, is called critical density, and thus the Universe has a density equal to the critical density, to measurement errors. Yet, if we count what we see actually in the Universe, we obtain a much lower density, approximately 1 % of the critical density: the density of the (predicted) Universe is bigger than the one that we observe. It is one of facets of the problem of the dark matter ".

Thus it is said " the density of the (predicted) Universe is bigger than the one that we observe ", 99 % times larger! Why do not say that the density of the Universe is the good one we observe, Einstein Relativity theory does not explain this observation, and thus that it would be false! No, physicists do not want to give up this beautiful mathematical theory and thus invent "dark matter" to explain the observation by the theory. The world of physics is gone crazy.

We have also seen that the theory leads to the Big Bang theory, which is not more valid than the dark matter or the dark energy, which are still necessary to add to make the set coherent. Let us resume now the text of Nainan Varghese. "We see well that in the mathematical domain from a set of hypotheses we can deduct a completely logical plan but which is connected with nothing concrete in the real physical world. At the end of a number of iterations consequences in comparisons with the real world, we can end in a construction certainly logical but completely absurd from the point of view of the reality ". This document in reference above expresses that we should not change hypotheses according to what we want to explain and that only one set of hypotheses must be used to explain all the observed phenomena's. It is not the case, in all the current theories on which the researchers are working in the field of "academic" physics. The "dissident" physicists propose new theories, with at the base have a philosophic ethics for the construction of a new theory, which is not to change the hypotheses along the way.

Chapter 13.

Gravitation and Inertia

Before going into the details of the new theories of the universe, this chapter is an introduction in order to address some basic issues.

Indeed, at the epoch of Newton it was distinguished two types of masses: inertial mass and gravitational mass.

As described by Isaac Newton's First Law of Motion, an object in motion should remain in motion in a straight line. The inertial mass of a body is its quality of resistance in any variation of its state of mechanical movement.

The gravitational mass comes from Newton's law on the attraction of bodies. Newton establishes that the strength of attraction of two bodies is proportional to the product of their mass divided by their square distance which separates them.

Further, experiments on gravity showing that the downfall of matter bodies is independent of their mass, Newton realized that gravitique mass and inertial mass are identical.

Newton could not explain the reason and Einstein neither because in its theory of relativity he set up this equality as a principle, the principle of equivalence. This is an elegant way of avoiding looking for a physical explanation of reality, which is that the slowness of a matter body and the strength of attraction of this one exercised on another one proceed of the same physical principle. This famous principle of equivalence prevented physicists from looking for the real cause.

To date, there is not in the sphere of academic theoretical physics, explanation of the physical phenomenon of inertia force of bodies. It is however one of the fundamental forces of our daily life! In addition, with a minimum of common sense, we see well that this explanation must be coherent of the physical phenomenon that produces gravitation.

In addition, as regards to the gravitational force, physicists speaks only about Einstein Relativity theory. For this one, the strength of attraction of two bodies comes from the curvature of space-time! In addition, how is it the curvature of space-time could explain inertia force? It is where the principle of equivalence saves Einstein Relativity theory.

Before Michelson & Morley experiment physicists thought that space was fill with an entity called "ether". Ether was supposed to be the support for photons propagation.

However, after the negative results of MM experiment, ether was no longer believed.

Some physicists now are supposing again that space is filled by ether and that it creates an energy field that can explain gravitation and inertia forces.

For gravitation force, since Einstein Relativity theory was accepted, physicists believe that gravitation force is due to bending of space-time coordinates and gravitation force is an attraction force between bodies.

In the new theories, gravitation is explained as a pressure force like the one Le Sage was proposing in the past.

Indeed, in 1758 the French physicist, Georges Louis Le Sage, proposed that gravitation force was not an attraction force but a pressure force! He was following Nicolas Fatio de Duillier who made the same hypothesis in 1690.

How pressures force can gives rise to an attraction force between two adjacent matter bodies?

Let suppose that a medium produces a pressure force on all directions on a particle of a matter body. Let suppose also that in between two close particles, the pressure force is lesser than in the outer directions for a reason to be explain later on. Then the two particles will tend to be closer and at the end, if there are no other interactions, they will merge. It is the way to explain that a pressure force can explain gravitation.

It is trivial to show that this gravitation force is exactly equal to Newton formulas, that is to say inversely proportional to square distance and proportional to product of masses.

What about inertia?

It is more complicated to explain in real terms, but we can say that the medium by the fact it acts as a pressure force towards a particle in all directions, whatever his motion is, and then its motion will stay forever. If we want to stop the motion of the particle, we will have to add a force to counteract the pressure force in the direction opposite to the displacement. However, we know by experience that the counteracting force must be proportional to the mass and the velocity of the matter body. We can see that the above explanation is not enough. We will have to details more of the new theories to see how in reality it can be explain.

Chapter 14.

New theories of the universe

Before dealing with the new theories of the universe, it is important to recall why it is necessary to give up with Einstein Relativity theory.

This theory states that our universe has the dimensions of "space-time", space which is empty, with the radio electric waves propagating without medium in this space and that however it is "bent" by matter.

Furthermore, the effect of "differential" speed between two reference frames leads to a slowing down of time and to a decrease of the lengths in the one who is faster. The simultaneity of the events in the universe is lost. This theory is not capable to explain the functioning of Galaxies;

relativists are obliged to introduce the notions of dark matter and dark energy into the universe!

We can dispute all these points, independently of the fact that all the experiences of validation of this relativist theory are refutable.

We showed that time has no physical reality as can have space, speed, and energy. Time is in fact only expression of movement of matter objects and has no physical reality. Why time seems to pass by in a direction and not in the other one while any equations of physics, which use the variable time (in the sense of a variable), are reversible with regard to time? The passing of time comes from the movement of matter objects either from an evolving process. The universe is everything but static, but in continual evolution due to energy transfer.

Relativists believe that time is not the same if one is at rest or in motion with high speed. They do not believe that the passage of time can be uniform in the universe. Here is a small mental exercise to explain that the "passage" of time is uniform for all the observers everywhere in the universe. Robert L. Henderson gives this example in the book in reference ([46]). Here is what it says. Two people leave the day of the summer solstice. The one stays on Earth when the other one leaves in a vehicle, which moves at a speed such as the theory of relativity leads to a slowing down of time inside its vessel, half the clock time being on Earth. The one who left the Earth is returning to the place where he left his accomplice after one year for the one who stayed on Earth. The one who stayed on Earth thus observed that the Earth made a complete turn around the sun. This is an absolutely true fact. For the person who comes back to earth after her journey to join his accomplice, his duration of journey should be of one half-year, then did he see that the Earth made an about-turn around the sun? No, it is not. Events cosmic or not are observed in an identical way in the entire universe, that we stayed on Earth or that we travel at high speed in space. The simultaneity of the temporal events is an indisputable fact.

Besides, to return to the slowing down of time, due to Einstein Relativity theory, and in particular of the aging of human beings, it have been shown that there is a contradiction to affirm on one side that "time" is dependent on the relative motion of reference frames and on the other side that all the laws of physics stays true in any reference frame. Living cells are subjects to the laws of physics, as aging of human beings, and then they are aging in the same ways in any reference frame!

Four new hypotheses for the functioning of the universe, try to give a rational explanation of all phenomena's observed in the universe.

To propose hypotheses on the functioning of the universe is a challenge, because they appeal to surprising notions in view of what is presented in Medias.

The following presentation is a bit more complicated given many researchers propose different hypotheses and they did not all reach the same level of development in their hypothesis.

All works are based on the fact that the space is constituted by a "medium" which "support" everything and is the "medium" for propagation of electromagnetic waves.

Besides the four proposals, which are going to be developed, there are other researchers publishing articles on these subjects, which go generally to the same spirit but without being supported still enough.

The most advanced hypotheses at present are proposed by the following authors (all these innovative hypotheses have to be credited to the following authors):

Robert L. Henderson ([47]),

Nainan K. Varghese: Matter ([48]).

Glen Borchart and Stephen J. Puetz ([49])

Thomas G. Lang ([50]).

The necessary simplifications brought in the following explanations, to present these hypotheses, can modify the spirit of the innovative ideas of the authors. The reader is thus encouraged to consult the documents in reference.

Robert L. Henderson

This theory developed by Robert L. Henderson goes back in the past because it was proposed by Nicolas Fatio de Duillier in 1690 then later by Georges-Louis Le Sage in 1748 ([51]).

In the book of R. L. Henderson, a big part is dedicated to show why the theory of Einstein seduced the scientific community and why according to him, it is not valid. We are not going to extend over this point which is already widely developed in the present book and with other arguments that those of Robert L. Henderson.

The starting point of this theory consists in saying that light, whether it is in the form of a wave or a particle, can move only in a propagation media, called "universal medium".

This theory makes the hypothesis that this medium, which fills the space, is composed of elementary particles in various states of extremely high velocity in random motion. Individual particles are supposed to have velocities varying in a random manner from zero to some as yet unknown upper limit. Particles are impinging upon each other with perfect elasticity.

To some extent it is like molecules in a gas, but the comparison stops there. Because the gravitation is a force, it is necessary that these particles have a mass. Robert L. Henderson calls this environment a field of energy (energy field).

Although this environment can seem extremely chaotic, it behaves as a force of pressure, due to the statistical effects.

How this medium can explain the gravitational force? Without getting into details given in the book, which cannot yet be proved, it is made the hypothesis that vis-à-vis a set of matter particles, the particles of the energy field, impacting on this set and crossing him, are reduced in number when they are going out on the other side. And so, if we consider that the set of matter particles is a sphere for example, then the particles of the energy field going inward are in smaller number than those going outward of the sphere. If another sphere of particles of material is close by, possessing the same properties, then the pressure force is reduced in between and both spheres are attracted.

How to explain the inertia force?

We did not find in the referenced document an explicit explanation of the inertia force.

We can just guess that when a sphere of matter particles wants to move, the particles of the energy field which impact on the sphere in opposite direction of the motion of the sphere, create a force counteracting this motion. Now if the sphere is in uniform motion, then the particles of the field of energy, which have a speed superior to the speed of the sphere, produce a force of the same intensity on all sides. This force maintains the speed of the sphere in its motion, which is the property of the inertia force observed at low speed.

We are not going to explain the various phenomena's observed in nature that the author describes in his book in reference, but simply to say that this pressure is the force, called in Quantum mechanics, "nuclear binding force", which is responsible for the stability of the components of atoms.

Reader can consult the document in reference to find explanations on the origin of matter, photons, radioactivity, electromagnetism with the electric and magnetic fields, etc...

Matter de Nainan K. Varghese

In this new hypothesis, only one postulation is used in the alternative concept on matter, proposed in the book 'MATTER (Re-examined)' and it is that 'substance is fundamental and matter provides substance to all real entities'. All physical phenomena's can be explained on this basis.

Here is a summary of this book given by the author.

As it is mention in the document, a free quantum of matter tends to reduce its existence into minimum number of spatial dimensions and grow into single spatial dimension. Although a quantum of matter has positive existence in all three spatial dimensions, due to inability of 3D beings to sense or measure distance below minimum tangible value, a free quantum of matter may be considered as '1D quantum of matter'. Under compression from all sides, matter-content of a quantum of matter may grow into all three spatial systems to make it a '3D quantum of matter'. 3D quanta of matter form all real (tangible and sensible) objects in universe.

Tendency of quanta of matter to grow into single spatial dimension helps them to form quadrilateral latticework-structures, called '2D energy-fields'. A 2D energy-field is a continuous latticework structure of quadrilaterals that extends to infinity in all directions, in its plane. 2D energy-fields in all possible planes in space, together, form universal medium.

Universal medium gathers and compresses free quanta of matter, available within gaps in it, to create disc-shaped 3D matter-cores of photons. Photons are the most basic 3D matter-particles and they constitute corpuscles of light or similar radiations. High-frequency photons, in various combinations, form primary 3D matter-particles, fundamental particles, and different types of atoms, molecules and macro bodies in universe. Distortions, in surrounding universal medium, sustain shapes and motions of 3D matter-cores of photons.

3D matter-cores of photons are created and sustained by universal medium. in order to maintain stability of both, universal medium and photon, 3D matter-core of a photon is spun about one of its diameter at spin speed proportional to its 3D matter-content and moved at highest possible (hence constant) linear speed by universal medium. These movements are accomplished by distortions in universal medium, formed about 3D matter-core. Inability of universal medium to move 3D matter beyond this speed limits linear speed of photon (light) and makes it constant with respect to surrounding universal medium.

Comments on inertia:

Due to their latticework structures and ability of self-stabilization, distortions introduced in 2D energy fields continue to transfer in straight-line direction, unless neutralized by distortions in opposite direction. During transfer of distortions, basic 3D matter-particles that happen to be within the region of distortions are also transferred along with moving distortions. Displacements of all basic 3D matter-particles in a macro body, together, result in motion of macro body.

An external force introduces distortions in 2D energy-fields about a 3D matter-body. Distortions in 2D energy-fields (universal medium) are work. It requires some time to introduce distortions in 2D energy-fields about a 3D matter-body and for their stabilization within the region of the matter-body (acceleration stage of the body). After stabilization, the distortions (work) will continue to move at constant linear speed in the same direction. As transfer of distortions (without displacements of structural elements in 2D energy-fields) is the cause of motion of a 3D matter-body, universal medium does not impede free motion of 3D matter-bodies (by drag or friction).

Ability of distortions (work about a 3D matter-body) to move at constant linear speed and the time delay required for their stabilization within the region of a matter-body, together, give rise to phenomenon of inertia.

Comments on gravitation:

Gravitation and gravitational attraction (gravity) are different phenomena. Universal medium is inherently under compression. Should there be a discontinuity in latticework-structures of 2D energy fields (even if it is due to presence of 3D matter), they have tendency to grow into the gap in their structures or they push at the 3D matter, present in the gap. This property of universal medium is gravitation. Gravitation is a static pressure. It creates basic 3D matter-particle and continues to act as long as the particles are in existence. Depending on the shape of surface, gravitation on a 3D matter-particle can be positive, negative or none. Gravitational pressure is enormously strong and it is manifested in the forms of 'natural forces'.

Magnitude of gravitation on a 3D matter-body corresponds to extent of universal medium that exerts the pressure. Extent of universal medium on the outer sides of two basic 3D matter-particles is always more than the extent between them. Higher pressure on their outer sides against lower pressure from in between, compels them to move towards each other. This tendency is understood as gravitational attraction (gravity). Gravitational attraction (gravity) is an apparent force (relatively a minor by-product of) resulting from separate gravitational actions on basic 3D matter-particles. Gravitational attraction (gravity) is active only between basic 3D matter-particles, whose disc-planes coincide at any instant. Each basic 3D matter-particle in one macro body gravitationally attracts every basic 3D matter-particle in another macro body.

3D matter-cores of constituent photons of a 3D matter-body move at the speed of light, in circular paths on imaginary surface of primary 3D matter-particles and spin in phase with each other. Therefore, it is rare for disc-plane of a photon in one macro body to coincide with disc-plane of a photon in another macro body and duration of such coincidence is extremely small. As a result, gravitational attraction between two macro bodies is very weak, compared to enormous strength of gravitation.

I shall refrain from describing all phenomena, explained by the author in his book.

Kindly refer to the book 'MATTER (Re-examined)' to find out more on origin of real entities, propagation of light, electromagnetism, cosmic bodies, etc.

Glen Borchardt et Stephen J. Puetz: Neomechanical Gravitation Theory.

It is suppose by the authors that what only really exist in the universe are matter and movement. The movement is not a part of the universe, but it is what does matter. The universal time is the movement of any things with regard to any things.

This theory distinguishes itself from the one of Le Sage and thus of Henderson by the fact that the particles of the ether are of varying size and can be small as possible. These particles collide constantly and when a smaller particle comes to the neighborhood of a bigger one, then as explained previously both particles are attracted to each other and it is what creates gravitation force and the cohesion of particles to form the particles of visible matter.

It is the slowing down of the particles of ether that get closer to each other that creates matter. To understand this phenomenon, the authors compare this formation with what takes place in a whirlwind. From a whirlwind, the surrounding particles are scattered with a lower density, then when the whirlwind builds up to itself then the density increase more and more until it reach a high density. It is what we see in the atmospheric tornados, where finally it is the surrounding energy, which focuses in a region.

The authors are considering that this principle of association of particles continues in all the scales in the universe.

For example, one of the first products of the complexity of ether is the atom of hydrogen, the simplest atom. It is the most present constituent in the universe; it represents 75 % of matter. The big clouds of atoms of hydrogen constitute basic elements to build the galaxies of stars and their planets.

One of the novelties brought by the authors is to consider that every element, whether it is the particles of ether or the elements of matter, are all different. What leads to the fact that everything, which is in the universe, heap of gas or atoms of the galaxies, are all distinct from each other.

The gravitational force and the inertia force are understandable in the same way as in the theories of Le Sage - Henderson.

The authors consider that the speed of propagation of light through ether depends on the density of this one. The density of ether being stronger at the level of the galaxies than in the intergalactic space, due to the effect of whirlwind, then the speed of light decreases when it goes to us and so would explain the "red-shift" of light coming from galaxies.

Unified Fluid Dynamic Theory of Physics

This theory was introduced by Thomas G. Lang in the 1950s and was published in the 2000s.

It makes the hypothesis that a fluid, the nature of which is not clarified by the author, constitutes the space.

Photons constitute the base of the construction of the universe, and are considered as a disturbance of the spatial fluid, itself allowing their propagation at a speed called the speed of light, which can vary locally.

Every photon is introduced by a local compression of the fluid under the shape of a sphere of very compressed fluid. Then this sphere is expanding and the internal pressure decreases until it become lower than the pressure of the fluid. It entails the recompression until it gains its size of origin and a displacement of a wavelength in the direction of the travel. As the motion of waves on the water, the spatial fluid does not move during the phases of expansion / recompression, only the photon moves.

Mass of photons resulting from the surrounding fluid, led a reduction of pressure in this fluid. Every photon is in the center of a region of the spatial fluid with a reduced pressure. This reduction of pressure is proportional to the mass of the photon and is supposed to vary in an inversely proportional way of the square distance to the center of the photon.

Matter is then constituted by photons of the way below. When two photons having the same orientation of their spin approach enough near to each other, then they are going to orbit to form a ring which is going to continue to pulse in the same way as the photons of origin, but be going to move with a speed of appropriate movement. This ring, if its spin turns contrary to watch hands, is called an electron and if its spin turns the other way, is called a positron.

A proton is formed by 919 positrons and by 918 electrons (these numbers result from the mass ratio between the mass of an electron / positron and that of a proton).

The creation of a neutron is more complex and the explanation exceeds the frame of this book. Interested reader can consult the literature on this subject. With these three components, combinations of these elements can create all atoms.

How to explain gravitation force?

We saw that a photon is in the center of a reduction of pressure of the spatial fluid, intensity of which is proportional to the mass and inversely proportional to the square distance. It is the same for components such as electrons, protons, and neutrons (positrons disappeared because they intervene in the formation of neutrons). Therefore, any mass approaching another mass is attracted towards it, what causes the gravitation.

How to explain inertia force of bodies in movement? We do not find in the document in reference an explicit explanation of the inertia force.

Others hypotheses

Another hypothesis is proposed by Mario Ludovico ([52]), but has not reached a stage of development yet being enough for explaining all the phenomena's in nature. The author leaves the principle that the ether is not constituted by particles but by a fluid, called a Cosmic Plenum, the properties of which he does not say clearly, and that this ether is surrounded by an absolute vacuum.

John-Erik Persson proposes ([53]), as Henderson whom he does not quote, that the ether consists of particles in movement, speed lower than that of the light, and of low mass. However, with regard to Le Sage and Henderson it is not clear whether it is the collision of the particles of ether, which produces a pressure on matter and thus the gravitation. He proposes that the inertia force is understandable by a wave, which adapts the movement of the ether surrounding to protect itself from the wind of ether.

Conclusion

Many physicists are thinking at present that the notion of time does not correspond to a real parameter, as space can be. Yet, all these physicists are thinking that Einstein Relativity theory is the true theory explaining the functioning of the universe. Then how on one side they can say that time does not exist and on the other side say that the universe is a space-time universe. It is a bit a schizophrenic attitude!

What is real in the universe are space and energy. Energy gives rise to velocity to matter bodies and/or gives rise to transformation of matter bodies by internal/external processing.

What we call an "instant of time" is a unique configuration of the universe where all matter bodies are defined by position and velocity state.

Going from one configuration to the other gives rise to what we call the "arrow of time".

If time does not really exist, then Einstein Relativity theory is no longer valid.

Nevertheless, the relativity principle remains a necessary principle, since all laws in nature have to be valid in any reference frame in motion or not.

Young-Sea Huang proposed a new theory of relativity, which uses a transformation which is based on displacement between two inertial reference frames in constant movement with each other's and which allows to preserve the validity of the laws of physics between these two inertial frames. This theory is compatible of the Heisenberg uncertainty principle and thus Quantum Mechanics. With this theory, time and space are of Newtonian nature and the synchronicity of events in the universe is preserved, what appears more in compliance with common sense.

It is said that Einstein Relativity theory has been validated by so many experiences that it cannot be put into question. Even the GPS is said to be operative thanks to Einstein Relativity theory. However, all this is untrue.

GPS is not working thanks to Einstein Relativity theory and many experiences that are said to be a validation of Einstein Relativity theory are untrue.

Relativistic unstable particles created in the atmosphere by cosmic rays (muons) are reaching the Earth ground because of the high speed of cosmic rays that can reach also Earth ground! David H. Frisch and James H. Smith measured almost the same quantity of cosmic rays (and muons) at Mount Washington and at ground level. Then there is no time dilation of muons lifetime going through the atmosphere.

It is very damaging that this experience is taught in every university in the world and every high school, and until now, nobody have seen that it is based on erroneous measurements.

New theories are proposed to explain the functioning of the universe, in particular to what the gravitational force and the inertia force might be attributed, by defining an "energy field" filling the entire universe. Each of these theories allows explaining phenomena's observed in nature and we see well the difficulty determining at the end, which one would be the right theory.

However, these new hypotheses for the functioning of the universe are proposed and nobody can see it! Then why the public has no knowledge of the works of these physicists?

In physics world, Einstein Relativity followers are prohibiting new ideas so that these new theories cannot emerge to the public and more than that, to the funding deciders.

In an article, Zbigniew Ozievicz ([54]) from Mexico University illustrates this prohibition. He simply recalls that in 1939, Landau & Lifshitz showed that in Einstein theory there is no concept of center of mass of a group of matter objects. In Newtonian physics, it is always possible to define a center of mass while it is not possible in

Einstein physics. The article was prohibited to be published by sensors guarantor of the dogma.

A group of independent physicists created an organization, Natural Philosophy Alliance (NPA), which allows all the physicists, whatever they are, to be able to present their works and a number of presentations are quoted in reference here. Of course we can find a very large number of different hypotheses for the functioning of the universe there, but the abundance is not a bad thing, given we teach at present a single theory which turns out to be untrue.

Academic science blame the world of the dissident scientists to not arriving at a consensus on their theories and thus these can only be false (quotation of David de Hilster ([55])).

Nowadays to criticize Einstein Relativity theory, Big Bang, black holes, the dark matter is a blasphemy.

Relativist followers will ask by what right one can carry such attacks towards people eminently recognized in their domain and often decorated with a Nobel Prize in Physics.

Any truth is good to say, even if it is to admit that we set up an icon of the physics by mistake.

How is it we have come to where we are today?

The first mistake was not to follow d'Alembert definition of acceleration, in which he uses duration ("dt") and not differential time. Then L. Euler gave a definition of acceleration using differential time, and then physicists became accustomed to use the variable time in equation as if it was a real entity of the universe.

The second mistake was due to Einstein in his Relativity theory in which time is thought as a real entity, while it is not. It is the movement of bodies and their transformation by processes, which are real, both caused by energy.

These errors led to a theory, which unfortunately made the consensus. Physicists do not thought out that another theory of relativity could exist and were not interested in physical explanations of the functioning of the universe.

It thus seems that there is a long way to go to achieve a comprehensive understanding of the universe, which is far from the mathematical theory of the universe supplied by the General Theory of Relativity.

Acknowledgements

Thanks to all the scientists quoted in the Reference Table. All these scientists, each in their domain, free of taboos, help expend our knowledge.

Reference Table

---

1 Dan Romalo. Absolute time ? Proceeding of the Natural Philosophy Alliance. 19th Annual Conference of the NPA. Volume 9. 25-28 july 2012.

2 Le temps est-il une illusion ? Numéro spécial « POUR LA SCIENCE ». N° 397 de Novembre 2010.

3 http://www.elishean.fr/le-temps-nexiste-pas/

4 FQXi community. The Nature of Time Essay Contest. http://fqxi.org/community/forum/category/10

5 According to 'MATTER (Re-examined)' http://gsjournal.net/Science-Journals/Research%20Papers-Astrophysics/Download/5260

6 R. Penrose. The Emperor's new mind, Penguin Press, 1991.

7 http://www.persee.fr/doc/rhs_0151-4105_1994_num_47_3_1214

8 http://stockage.univ-brest.fr/~acolindv/telecharger/meca_physique/partie_2.pdf

9 https://fr.wikipedia.org/wiki/Temps

10 https://fr.wikipedia.org/wiki/Histoire_de_la_mesure_du_temps

[11] Peter Kohut. E=mc2 and Einstein's failure. Proceeding of the Natural Philosophy Alliance. 19th Annual Conference of the NPA. Volume 9. 25-28 july 2012.

12 https://www.herodote.net/532_a_726-synthese-27.php

13 Ciel & espace. L'univers de l'association française d'astronomie. Gravitation, la magie qui gouverne l'univers. Juillet/Août 2017.

14 https://www.quora.com/What-is-the-difference-between-general-and-special-relativity

15 http://www.sciences.ch/htmlfr/cosmologie/cosmorelativisteres01.php

16 Young-Sea Huang. An alternative to relativistic transformation of special relativity based on the first principles. arXiv:0812.5029v1. 30 Dec 2008.

17 Young-Sea Huang. Relativistic quantum mechanics and relativistic quantum statistics based upon a novel perspective on relativistic transformation. Research gate.

18 G Peter H. Dana. Global Positioning System Overview.

19 Neil Ashby. Relativity in the Global Positioning System. http://relativity.livingreviews.org/Articles/lrr-2003-1/download/lrr-2003-1Color.pdf

20 Barry Springer. Does GPG Navigation Rely upon Einstein's Relativity? Proceedings of the Natural Philosophy Alliance. 20th Annual Conference of the NPA, 10-13 July, 2013. College Park. Maryland

21 Jean Marc Piéplu. GPS et Galileo. Systèmes de navigation par satellites. https://books.google.fr/books?isbn=2212047363

22 DAVID H. FRISCH ET JAMES H. SMITH. Mesure de la dilatation relativiste du temps utilisant les mésons-μ. Science Teaching Center et Departement of Physics, MIT, Cambridge, Massachusetts.

23 http://www.ph.surrey.ac.uk/satellites/main/tutorial2_1.html

24 https://cds.cern.ch/record/557167/files/p41.pdf

25 Francis J. M. Farley. Muon g − 2 and Tests of Relativity. Energy and Climate Change Division, Engineering and the Environment, Southampton University, Highfield, Southampton, SO17 1BJ, England, UK.

26 J. Bailey, W. Bartl, G. von Bochmann, R.C.A. Brown, F.J.M. Farley, M. Giesch, H. Jöstlein, S. van der Meer, E. Picasso and R.W. Williams, Nuovo Cimento A 9, 369 (1972).

27 Young-Sea-Huang. Einstein's Relativistic Time-Dilation: A Critical Analysis and a Suggested Experiment Helv. Phys. Acta 66, 346 (1993).

28 Young-Sea Huang. A Comment on "The Tests of Relativistic Time Dilation in the CERN Muon Storage Ring".

29 Francis J.M. Farley. The magnetic moment of the muon worries theorists. Europhysics News (2001) Vol. 32 No. 5

30 Roger Rydin https://www.youtube.com/watch?v=rv5GJCkI-bk&feature=youtu.be&utm_source=Natural+Philosophy+Alliance+Newsletter&utm_campaign=ce0a63c17d-NPA_Update_Conference_Presentation_Videos7_8_2015&utm_medium=email&utm_term=0_5c1849dfda-ce0a63c17d-157622721

31 Nainan Varghese. Precession of Mercury's orbit. http://www.science20.com/matter/blog/precession_mercury%E2%80%99s_orbit.

32 Edward H. Dowdye, Jr. Gravitational Light Bending History is Severely Impact-Parameter Dependent. Proceeding of the Natural Philosophy Alliance. 19th Annual Conference of the NPA. Volume 9. 25-28 july 2012.

33 Young-Sea_Huang. Einstein's Relativistic Time-Dilation: A Critical Analysis and a Suggested Experiment Helv. Phys. Acta 66, 346 (1993).

34 https://fr.wikipedia.org/wiki/Big_Bang.

35 Lyndon Ashmore. New Tired Light Correctly Predicts the Redshift of the CorBor Galaxy Cluster. Proceedings of the Natural Philosophy Alliance. 20 th Annual Conference of the NPA. 10 – 13 July 2013.

36 Stephen J. Crothers. Proof of no « Black Hole » Binary in Nova Scorpii.. Proceeding of the Natural Philosophy Alliance. 19th Annual Conference of the NPA. Volume 9. 25-28 july 2012.

37
https://en.wikipedia.org/wiki/Michelson%E2%80%93Morley_experi
ment

38 https://en.wikipedia.org/wiki/Fizeau_experiment

39 Jean David. http://jean.david.free.fr/michelson-morley.pdf

40 https://www.omicsonline.org/open-access/optical-fizeau-
experiment-with-moving-water-is-explained-withoutfresnels-
hypothesis-and-contradicts-special-relativity-2090-0902-
1000207.php?aid=86879

41 https://www.bibnum.education.fr/sites/default/files/Sagnac-
analyse.pdf

42 Peter Sujak . On the General Reality of Gravity, as well as other
forces in nature and the creation of material particles and force fields
in the universe., Hradinska 60, 10100 Prague, Czech Republik.
Proceedings of the Natural Philosophy Alliance. 20th Annual
Conference of the NPA

43 https://fr.wikipedia.org/wiki/M%C3%A9canique_quantique

44 https://arxiv.org/pdf/1301.1069v1.pdf

45 Nainan K. Varghese, basic assumption in physics.
matterdoc@gmail.com, http://www.matterdoc.info

46 Robert L. Henderson. Einstein and The Emperor's New-Clothes
Syndrome.

47 Robert L. Henderson. Einstein and the-Emperor's new clothes syndrome. The exposé of a charlatan.

48 Nainan K. Varghese. Matter, an alternative concept, http://www.matterdoc.info/

49 Neo mechanical Gravitation Theory. Proceeding of the Natural Philosophy Alliance. 19th Annual Conference of the NPA. Volume 9. 25-28 july 2012

50 Tomas G. Lang. Unified fluid-based theory of physics. Proceeding of the Natural Philosophy Alliance. 20th Annual Conference of the NPA. 10-13 july 2013.

51 https://en.wikipedia.org/wiki/Le_Sage's_theory_of_gravitation

52 Gravitational Vortexes of Cosmic Plenum. Proceeding of the Natural Philosophy Alliance. 19th Annual Conference of the NPA. Volume 9. 25-28 july 2012.

53 John-Erik Person. The Falling Ether. Proceeding of the Natural Philosophy Alliance. 20th Annual Conference of the NPA. 10-13 july 2013

54 Zbigniew *Oziewicz. Are Peer Reviewers Guardians of the Truth ? Proceeding of the Natural Philosophy Alliance. 20th Annual Conference of the NPA. 10-13 july 2013.*

55 David *de Hilster. Consensus in science is wrong. Proceeding of the Natural Philosophy Alliance. 19*[th]